Observability with Grafana

Monitor, control, and visualize your Kubernetes
and cloud platforms using the LGTM stack

Rob Chapman

Peter Holmes

<packt>

BIRMINGHAM—MUMBAI

Observability with Grafana

Group Product Manager: Preet Ahuja
Publishing Product Manager: Surbhi Suman
Book Project Manager: Ashwini Gowda
Senior Editor: Shruti Menon
Technical Editor: Nithik Cheruvakodan
Copy Editor: Safis Editing
Proofreader: Safis Editing
Indexer: Tejal Daruwale Soni
Production Designer: Ponraj Dhandapani
DevRel Marketing Coordinator: Rohan Dobhal
Senior DevRel Marketing Coordinator: Linda Pearlson

First published: December 2023

Production reference: 1141223

Published by Packt Publishing Ltd.

Grosvenor House
11 St Paul's Square
Birmingham
B3 1RB, UK

ISBN 978-1-80324-800-4

www.packtpub.com

To all my children, for making me want to be more. To Heather, for making that possible.

– Rob Chapman

For every little moment that brought me to this point.

– Peter Holmes

Contributors

About the authors

Rob Chapman is a creative IT engineer and founder at The Melt Cafe, with two decades of experience in the full application life cycle. Working over the years for companies such as the Environment Agency, BT Global Services, Microsoft, and Grafana, Rob has built a wealth of experience on large complex systems. More than anything, Rob loves saving energy, time, and money and has a track record for bringing production-related concerns forward so that they are addressed earlier in the development cycle, when they are cheaper and easier to solve. In his spare time, Rob is a Scout leader, and he enjoys hiking, climbing, and, most of all, spending time with his family and six children.

A special thanks to Peter, my co-author/business partner, and to our two reviewers, Flick and Brad – we did this!

Heather, my best friend – thank you for believing in me and giving me space to grow.

Thank you to my friend and coach, Sam, for guiding me to be all I can be and not being afraid to show that to the world.

Last but not least, thanks to Phil for the sanctuary over the years – you kept me sane.

Peter Holmes is a senior engineer with a deep interest in digital systems and how to use them to solve problems. With over 16 years of experience, he has worked in various roles in operations. Working at organizations such as Boots UK, Fujitsu Services, Anaplan, Thomson Reuters, and the NHS, he has experience in complex transformational projects, site reliability engineering, platform engineering, and leadership. Peter has a history of taking time to understand the customer and ensuring Day-2+ operations are as smooth and cost-effective as possible.

A special thanks to Rob, my co-author, for bringing me along on this writing journey, and to our reviewers.

Rania, my wife – thank you for helping me stay sane while writing this book.

About the reviewers

Felicity (Flick) Ratcliffe has almost 20 years of experience in IT operations. Starting her career in technical support for an internet service provider, Flick has used her analytical and problem-solving superpowers to grow her career. She has worked for various companies over the past two decades in frontline operational positions, such as systems administrator, site reliability engineer, and now, for Post Office Ltd in the UK, as a cloud platform engineer. Continually developing her strong interest in emerging technologies, Flick has become a specialist in the area of observability over the past few years and champions the cause for OpenTelemetry within the organizations she works at or comes into contact with.

I have done so much with my career in IT, thanks to my colleagues, past and present. Thank you for your friendship and mentorship. You encouraged me to develop my aptitude for technology, and I cannot imagine myself in any other type of work now.

Bradley Pettit has over 10 years of experience in the technology industry. His expertise spans a range of roles, including hands-on engineering and solution architecture. Bradley excels in addressing complex technical challenges, thanks to his strong foundation in platform and systems engineering, automation, and DevOps practices. Recently, Bradley has specialized in observability, working as a senior solutions architect at Grafana Labs. He is a highly analytical, dedicated, and results-oriented professional. Bradley's customer-centric delivery approach empowers organizations and the *O11y* community to achieve transformative outcomes.

Table of Contents

Preface   xv

Part 1: Get Started with Grafana and Observability

1

Introducing Observability and the Grafana Stack   3

Observability in a nutshell	4	Pelé Product	16
Case study – A ship passing through the		Masha Manager	18
Panama Canal	5	**Introducing the Grafana stack**	**19**
Telemetry types and technologies	**6**	The core Grafana stack	19
Metrics	7	Grafana Enterprise plugins	21
Logs	7	Grafana incident response and management	21
Distributed traces	8	Other Grafana tools	21
Other telemetry types	10	**Alternatives to the Grafana stack**	**22**
Introducing the user personas of		Data collection	23
observers	**11**	Data storage, processing, and visualization	23
Diego Developer	13	**Deploying the Grafana stack**	**24**
Ophelia Operator	14	**Summary**	**25**
Steven Service	15		

2

Instrumenting Applications and Infrastructure   27

Common log formats	28	Structured, semi-structured, and	
		unstructured logging	28

Sample log formats   30

Exploring metric types and best practices   34

Metric types   35
Comparing metric types   36
Metric protocols   38
Best practices for implementing metrics   39

Tracing protocols and best practices   40

Spans and traces   40
Tracing protocols   41

Best practices for setting up distributed tracing 42

Using libraries to instrument efficiently 43

Popular libraries for different programming
languages   44

Infrastructure data technologies   45

Common infrastructure components   45
Common standards for infrastructure
components   47

Summary   48

3

Setting Up a Learning Environment with Demo Applications   49

Technical requirements   49
Introducing Grafana Cloud   50

Setting up an account   50
Exploring the Grafana Cloud Portal   53
Exploring the Grafana instance   56

Installing the prerequisite tools   60

Installing WSL2   61
Installing Homebrew   61
Installing container orchestration tools   62
Installing a single-node Kubernetes cluster   63
Installing Helm   63

Installing the OpenTelemetry Demo application   64

Setting up access credentials   64
Downloading the repository and adding
credentials and endpoints   64
Installing the OpenTelemetry Collector   66

Installing the OpenTelemetry demo application 67

Exploring telemetry from the demo application   68

Logs in Loki   69
Metrics in Prometheus/Mimir   72
Traces in Tempo   73
Adding your own applications   76

Troubleshooting your OpenTelemetry Demo installation   76

Checking Grafana credentials   76
Reading logs from the OpenTelemetry Collector 77
Debugging logs from the OpenTelemetry
Collector   77

Summary   78

Part 2: Implement Telemetry in Grafana

4

Looking at Logs with Grafana Loki   81

Technical requirements	81	Log stream selector	88
Updating the OpenTelemetry demo		Log pipeline	88
application	82	Exploring LogQL metric queries	95
Introducing Loki	82	**Exploring Loki's architecture**	**99**
Understanding LogQL	84	**Tips, tricks, and best practices**	**101**
LogQL query builder	84	**Summary**	**103**
An overview of LogQL features	86		

5

Monitoring with Metrics Using Grafana Mimir and Prometheus   105

Technical requirements	105	Prometheus	121
Updating the OpenTelemetry demo		SNMP	121
application	106	**Understanding data storage**	
Introducing PromQL	106	**architectures**	**121**
An overview of PromQL features	108	Graphite architecture	122
Writing PromQL	114	Prometheus architecture	122
		Mimir architecture	123
Exploring data collection and metric		**Using exemplars in Grafana**	**125**
protocols	**119**	**Summary**	**127**
StatsD and DogStatsD	120		
OTLP	120		

6

Tracing Technicalities with Grafana Tempo   129

Technical requirements	129	**Introducing Tempo and the TraceQL**	
Updating the OpenTelemetry Demo		**query language**	**130**
application	129	Exploring the Tempo features	131
		Exploring the Tempo Query language	135

Pivoting between data types   138

Exploring tracing protocols   **139**

What are the main tracing protocols? 139

Context propagation   140

Understanding the Tempo
architecture   143

Summary   145

7

Interrogating Infrastructure with Kubernetes, AWS, GCP, and Azure 147

Technical requirements   **147**

Monitoring Kubernetes using Grafana 148

Kubernetes Attributes Processor   148

Kubeletstats Receiver   150

Filelog Receiver   151

Kubernetes Cluster Receiver   152

Kubernetes Object Receiver   152

Prometheus Receiver   153

Host Metrics Receiver   154

**Visualizing AWS telemetry with
Grafana Cloud**   **155**

Amazon CloudWatch data source   155

Exploring AWS integration   157

Monitoring GCP using Grafana   **162**

Configuring the data source   162

Google Cloud Monitoring query editor 163

Google Cloud Monitoring dashboards   165

Monitoring Azure using Grafana   **166**

Configuring the data source   166

Using the Azure Monitor query editor 167

Using Azure Monitor dashboards   170

Best practices and approaches   **171**

Summary   **172**

Part 3: Grafana in Practice

8

Displaying Data with Dashboards 175

Technical requirements   **175**

Creating your first dashboard   **176**

Developing your dashboard further **180**

Using visualizations in Grafana **183**

Developing a dashboard purpose   **185**

Advanced dashboard techniques   **186**

Managing and organizing dashboards 189

Case study – an overall system view **191**

Summary   **195**

9

Managing Incidents Using Alerts   197

Technical requirements	198	Groups and admin	213
Being alerted versus being alarmed	198	**Grafana OnCall**	**214**
Before an incident	199	Alert groups	214
During an incident	202	Inbound integrations	215
After an incident	207	Templating	217
		Escalation chains	218
Writing great alerts using SLIs and SLOs	208	Outbound integrations	219
		Schedules	221
Grafana Alerting	209		
Alert rules	210	**Grafana Incident**	**222**
Contact points, notification policies, and silences	213	**Summary**	**224**

10

Automation with Infrastructure as Code   225

Technical requirements	225	Getting to grips with the Grafana API	236
Benefits of automating Grafana	226	Exploring the Grafana Cloud API	236
Introducing the components of observability systems	226	Using Terraform and Ansible for Grafana Cloud	237
		Exploring the Grafana API	239
Automating collection infrastructure with Helm or Ansible	228	Managing dashboards and alerts with Terraform or Ansible	240
Automating the installation of the OpenTelemetry Collector	228	**Summary**	**242**
Automating the installation of Grafana Agent	235		

11

Architecting an Observability Platform   243

Architecting your observability platform	243	Management and automation	253
Defining a data architecture	244	Developing a proof of concept	254
Establishing system architecture	246	Containerization and virtualization	254

Data production tools	255	Sending telemetry to other consumers	258
Setting the right access levels	256	Summary	259

Part 4: Advanced Applications and Best Practices of Grafana

12

Real User Monitoring with Grafana   263

Introducing RUM	263	Pivoting from frontend to backend data	271
Setting up Grafana Frontend Observability	265	Enhancements and custom configurations	273
Exploring Web Vitals	269	Summary	274

13

Application Performance with Grafana Pyroscope and k6   275

Using Pyroscope for continuous profiling	276	A brief overview of k6	285
A brief overview of Pyroscope	276	Writing a test using checks	286
Searching Pyroscope data	276	Writing a test using thresholds	286
Continuous profiling client configuration	279	Adding scenarios to a test to run at scale	287
Understanding the Pyroscope architecture	282	Test life cycle	288
		Installing and running k6	288
Using k6 for load testing	283	Summary	290

14

Supporting DevOps Processes with Observability   291

Introducing the DevOps life cycle	292	Release	295
Using Grafana for fast feedback during the development life cycle	294	Deploy	296
		Operate	298
Code	294	Monitor	298
Test	295	Plan	300

Using Grafana to monitor
infrastructure and platforms   300
Observability platforms   300
CI platforms   301

CD platforms   302
Resource platforms   303
Security platforms   303
Summary   304

15

Troubleshooting, Implementing Best Practices, and More with Grafana
305

Best practices and troubleshooting for
data collection   305
Preparing for data collection   306
Data collection decisions   306
Debugging collector   307

Best practices and troubleshooting for
the Grafana stack   308

Preparing the Grafana stack   308
Grafana stack decisions   309
Debugging Grafana   309

Avoiding pitfalls of observability   311

Future trends in application
monitoring   312
Summary   313

Index   315

Other Books You May Enjoy   330

Preface

Hello and welcome! *Observability with Grafana* is a book about the tools offered by Grafana Labs for observability and monitoring. Grafana Labs is an industry-leading provider of open source tools to collect, store, and visualize data collected from IT systems. This book is primarily aimed toward IT engineers who will interact with these systems, whatever discipline they work in.

We have written this book as we have seen some common problems across organizations:

- Systems that were designed without a strategy for scaling are being pushed to handle additional data load or teams using the system

- Operational costs are not being attributable correctly in the organization, leading to poor cost analysis and management

- Incident management processes that treat the humans involved as robots without sleep schedules or parasympathetic nervous systems

In this book, we will use the OpenTelemetry Demo application to simulate a real-world environment and send the collected data to a free Grafana Cloud account that we will create. This will guide you through the Grafana tools for collecting telemetry and also give you hands-on experience using the administration and support tools offered by Grafana. This approach will teach you how to run the Grafana tools in a way so that anyone can experiment and learn independently.

This is an exciting time for Grafana, identified as a visionary in the *2023 Gartner Magic Quadrant for Observability* (https://www.gartner.com/en/documents/4500499). They recently delivered change in two trending areas:

- **Cost reduction**: This has seen Grafana as the first vendor in the observability space to release tools that not only help you understand your costs but also reduce them.

- **Artificial intelligence** (**AI**): Grafana has introduced generative AI tools that assist daily operations in simple yet effective ways – for example, writing an incident summary automatically. Grafana Labs also recently purchased *Asserts.ai* to simplify root cause analysis and accelerate problem detection.

We hope you enjoy learning some new things with us and have fun doing it!

Who this book is for

IT engineers, support teams, and leaders can gain practical insights into bringing the huge power of an observability platform to their organization. The book will focus on engineers in disciplines such as the following:

- **Software development**: Learn how to quickly instrument applications and produce great visualizations enabling applications to be easily supported

- **Operational teams** (DevOps, Operations, Site Reliability, Platform, or Infrastructure): Learn to manage an observability platform or other key infrastructure platform and how to manage such platforms in the same way as any other application

- **Support teams**: Learn how to work closely with development and operational teams to have great visualizations and alerting in place to quickly respond to customers' needs and IT incidents

This book will also clearly establish the role of leadership in incident management, cost management, and establishing an accurate data model for this powerful dataset.

What this book covers

Chapter 1, Introducing Observability and the Grafana Stack, provides an introduction to the Grafana product stack in relation to observability as a whole. You will learn about the target audiences and how that impacts your design. We will take a look at the roadmap for observability tooling and how Grafana compares to alternative solutions. We will explore architectural deployment models, from self-hosted to cloud offerings. Inevitably, you will have answers to the question "Why choose Grafana?".

Chapter 2, Instrumenting Applications and Infrastructure, takes you through the common protocols and best practices for each telemetry type at a high level. You will be introduced to widely used libraries for multiple programming languages that make instrumenting an application simple. Common protocols and strategies for collecting data from infrastructural components will also be discussed. This chapter provides a high-level overview of the technology space and aims to be valuable for quick reference.

Chapter 3, Setting Up a Learning Environment with Demo Applications, explains how to install and set up a learning environment that will support you through later sections of the book. You will also learn how to explore the telemetry produced by the demo app and add monitoring for your own service.

Chapter 4, Looking at Logs with, Loki, takes you through working examples to understand LogQL. You will then be introduced to common log formats, and their benefits and drawbacks. Finally, you will be taken through the important architectural designs of Loki, and best practices when working with it.

Chapter 5, Monitoring with Metrics Using Grafana Mimir and Prometheus, discusses working examples to understand PromQL with real data. Detailed information about the different metric protocols will be discussed. Finally, you will be taken through important architectural designs backing Mimir, Prometheus, and Graphite that guide best practices when working with the tools.

Chapter 6, Tracing Technicalities with Grafana Tempo, shows you working examples to understand TraceQL with real data. Detailed information about the different tracing protocols will be discussed. Finally, you will be taken through the important architectural designs of Tempo, and best practices when working with it.

Chapter 7, Interrogating Infrastructure with Kubernetes, AWS, GCP, and Azure, describes the setup and configuration used to capture telemetry from infrastructure. You will learn about the different options available for Kubernetes. Additionally, you will investigate the main plugins that allow Grafana to query data from cloud vendors such as AWS, GCP, and Azure. You will look at solutions to handle large volumes of telemetry where direct connections are not scalable. The chapter will also cover options for filtering and selecting telemetry data before it gets to Grafana for security and cost optimization.

Chapter 8, Displaying Data with Dashboards, explains how you can set up your first dashboard in the Grafana UI. You will also learn how to present your telemetry data in an effective and meaningful way. The chapter will also teach you how to manage your Grafana dashboards to be organized and secure.

Chapter 9, Managing Incidents Using Alerts, describes how to set up your first Grafana alert with Alert Manager. You will learn how to design an alert strategy that prioritizes business-critical alerts over ordinary notifications. Additionally, you will learn about alert notification policies, different delivery methods, and what to look for.

Chapter 10, Automation with Infrastructure as Code, gives you the tools and approaches to automate parts of your Grafana stack deployments while introducing standards and quality checks. You will gain a deep dive into the Grafana API, working with Terraform, and how to protect changes with validation.

Chapter 11, Architecting an Observability Platform, will show those of you who are responsible for offering an efficient and easy-to-use observability platform how you can structure your platform so you can delight your internal customers. In the environment we operate in, it is vital to offer these platform services as quickly and efficiently as possible, so more time can be dedicated to the production of customer-facing products. This chapter aims to build on the ideas already covered to get you up and running quickly.

Chapter 12, Real User Monitoring with Grafana, introduces you to frontend application observability, using Grafana Faro and Grafana Cloud Frontend Observability for **real user monitoring** (**RUM**). This chapter will discuss instrumenting your frontend browser applications. You will learn how to capture frontend telemetry and link this with backend telemetry for full stack observability.

Chapter 13, Application Performance with Grafana Pyroscope and k6, introduces you to application performance and profiling using Grafana Pyroscope and k6. You will obtain a high-level overview that discusses the various aspects of k6 for smoke, spike, stress, and soak tests, as well as using Pyroscope to continuously profile an application both in production and test environments.

Chapter 14, Supporting DevOps Processes with Observability, takes you through DevOps processes and how they can be supercharged with observability with Grafana. You will learn how the Grafana stack can be used in the development stages to speed up the feedback loop for engineers. You will understand how to prepare engineers to operate the product in production. Finally, You will learn when and how to implement CLI and automation tools to enhance the development workflow.

Chapter 15, Troubleshooting, Implementing Best Practices, and More with Grafana, closes the book by taking you through best practices when working with Grafana in production. You will also learn some valuable troubleshooting tips to support you with high-traffic systems in day-to-day operations. You will also learn about additional considerations for your telemetry data with security and business intelligence.

To get the most out of this book

The following table presents the operating system requirements for the software that will be used in this book:

Software/hardware covered in the book	Operating system requirements
Kubernetes v1.26	Either Windows, macOS, or Linux with dual CPU and 4 GB RAM
Docker v23	Either Windows, macOS, or Linux with dual CPU and 4 GB RAM

If you are using the digital version of this book, we advise you to type the code yourself or access the code from the book's GitHub repository (a link is available in the next section). Doing so will help you avoid any potential errors related to the copying and pasting of code.

Download the example code files

You can download the example code files for this book from GitHub at `https://github.com/PacktPublishing/Observability-with-Grafana`. If there's an update to the code, it will be updated in the GitHub repository.

We also have other code bundles from our rich catalog of books and videos available at `https://github.com/PacktPublishing/`. Check them out!

Code in Action

The Code in Action videos for this book can be viewed at `https://packt.link/v59Jp`.

Conventions used

There are a number of text conventions used throughout this book.

`Code in text`: Indicates code words in text, database table names, folder names, filenames, file extensions, pathnames, dummy URLs, user input, and Twitter handles. Here is an example: "The function used to get this information is the `rate()` function."

A block of code is set as follows:

```
histogram_quantile(
    0.95, sum(
        rate(
            http_server_duration_milliseconds_bucket{}[$__rate_
interval])
        ) by (le)
    )
```

Any command-line input or output is written as follows:

```
$ helm upgrade owg open-telemetry/opentelemetry-collector -f OTEL-
Collector.yaml
```

Bold: Indicates a new term, an important word, or words that you see onscreen. For instance, words in menus or dialog boxes appear in **bold**. Here is an example: "From the dashboard's **Settings** screen, you can add or remove tags from individual dashboards."

> **Tips or important notes**
> Appear like this.

Get in touch

Feedback from our readers is always welcome.

General feedback: If you have questions about any aspect of this book, email us at `customercare@packtpub.com` and mention the book title in the subject of your message.

Errata: Although we have taken every care to ensure the accuracy of our content, mistakes do happen. If you have found a mistake in this book, we would be grateful if you would report this to us. Please visit `www.packtpub.com/support/errata` and fill in the form.

Piracy: If you come across any illegal copies of our works in any form on the internet, we would be grateful if you would provide us with the location address or website name. Please contact us at `copyright@packt.com` with a link to the material.

If you are interested in becoming an author: If there is a topic that you have expertise in and you are interested in either writing or contributing to a book, please visit `authors.packtpub.com`.

Share Your Thoughts

Once you've read *Observability with Grafana*, we'd love to hear your thoughts! Scan the QR code below to go straight to the Amazon review page for this book and share your feedback.

https://packt.link/r/1803248009

Your review is important to us and the tech community and will help us make sure we're delivering excellent quality content.

Download a free PDF copy of this book

Thanks for purchasing this book!

Do you like to read on the go but are unable to carry your print books everywhere?

Is your eBook purchase not compatible with the device of your choice?

Don't worry, now with every Packt book you get a DRM-free PDF version of that book at no cost.

Read anywhere, any place, on any device. Search, copy, and paste code from your favorite technical books directly into your application.

The perks don't stop there, you can get exclusive access to discounts, newsletters, and great free content in your inbox daily

Follow these simple steps to get the benefits:

1. Scan the QR code or visit the link below

https://packt.link/free-ebook/9781803248004

2. Submit your proof of purchase
3. That's it! We'll send your free PDF and other benefits to your email directly

Part 1: Get Started with Grafana and Observability

In this part of the book, you will get an introduction to Grafana and observability. You will learn about the producers and consumers of telemetry data. You will explore how to instrument applications and infrastructure. Then, you will set up a learning environment that will be enhanced throughout the chapters to provide comprehensive examples of all parts of observability with Grafana.

This part has the following chapters:

- *Chapter 1, Introducing Observability and the Grafana Stack*
- *Chapter 2, Instrumenting Applications and Infrastructure*
- *Chapter 3, Setting Up a Learning Environment with Demo Applications*

1

Introducing Observability and the Grafana Stack

The modern computer systems we work with have moved from the realm of complicated into the realm of complex, where the number of interacting variables make them ultimately unknowable and uncontrollable. We are using the terms *complicated* and *complex* as per system theory. A complicated system, like an engine, has clear causal relationships between components. A complex system, such as the flowing of traffic in a city, shows emergent behavior from the interactions of its components.

With the average cost of downtime estimated to be $9,000 per minute by Ponemon Institute in 2016, this complexity can cause significant financial loss if organizations do not take steps to manage this risk. **Observability** offers a way to mitigate these risks, but making systems observable comes with its own financial risks if implemented poorly or without a clear business goal.

In this book, we will give you a good understanding of what observability is and who the customers who might use it are. We will explore how to use the tools available from Grafana Labs to gain visibility of your organization. These tools include Loki, Prometheus, Mimir, Tempo, Frontend Observability, Pyroscope, and k6. You will learn how to use **Service Level Indicators** (**SLIs**) and **Service Level Objectives** (**SLOs**) to obtain clear transparent signals of when a service is operating correctly, and how to use the Grafana **incident response tools** to handle incidents. Finally, you will learn about managing their observability platform using **automation tools** such as Ansible, Terraform, and Helm.

This chapter aims to introduce observability to all audiences, using examples outside of the computing world. We'll introduce the types of **telemetry** used by observability tools, which will give you an overview of how to use them to quickly understand the state of your services. The various **personas** who might use observability systems will be outlined so that you can explore complex ideas later with a clear grounding on who will benefit from their correct implementation. Finally, we'll investigate Grafana's **Loki, Grafana, Tempo, Mimir** (**LGTM**) stack, how to deploy it, and what alternatives exist.

In this chapter, we're going to cover the following main topics:

- Observability in a nutshell
- Telemetry types and technologies
- Understanding the customers of observability
- Introducing the Grafana stack
- Alternatives to the Grafana stack
- Deploying the Grafana stack

Observability in a nutshell

The term **observability** is borrowed from control theory. It's common to use the term interchangeably with the term **monitoring** in IT systems, as the concepts are closely related. Monitoring is the ability to raise an alarm when something is wrong, while observability is the ability to understand a system and determine whether something is wrong, and why.

Control theory was formalized in the late 1800s on the topic of centrifugal governors in steam engines. This diagram shows a simplified view of such a system:

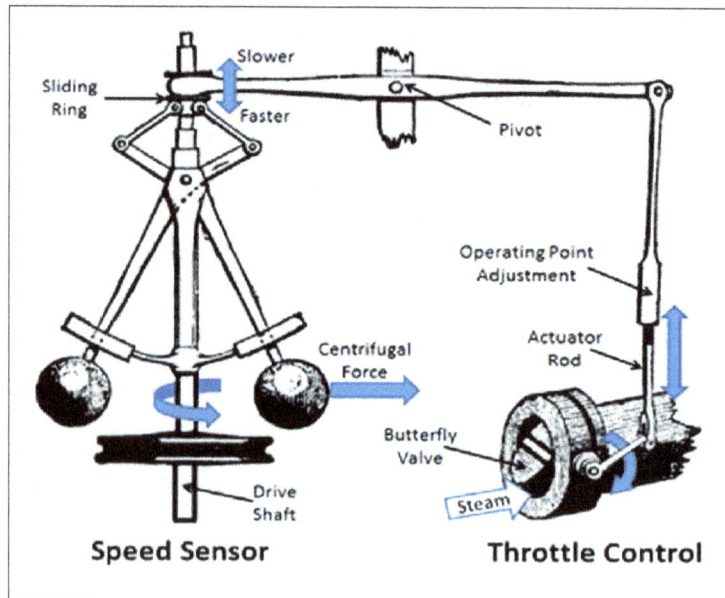

Figure 1.1 – James Watt's steam engine flyweight governor (source: https://www.mpoweruk.com)

Steam engines use a boiler to boil water in a pressure vessel. The steam pushes a piston backward and forward, which converts heat energy to reciprocal energy. In steam engines that use centrifugal governors, this reciprocal energy is converted to rotational energy via a wheel connected to a piston. The centrifugal governor provides a physical link backward through the system to the throttle. This means that the speed of rotation controls the throttle, which, in turn, controls the speed of rotation. Physically, this is observed by the balls on the governor flying outward and dropping inward until the system reaches equilibrium.

Monitoring defines the metrics or events that are of interest in advance. For instance, the governor measures the pre-defined metric of drive shaft revolutions. The controllability of the throttle is then provided by the pivot and actuator rod assembly. Assuming the actuator rod is adjusted correctly, the governor should control the throttle from fully open to fully closed.

In contrast, observability is achieved by allowing the internal state of the system to be inferred from its external outputs. If the operating point adjustment is incorrectly set, the governor may spin too fast or too slowly, rendering the throttle control ineffective. A governor spinning too fast or too slowly could also indicate that the sliding ring is stuck in place and needs oiling. Importantly, this insight can be gained without defining in advance what too fast or too slow means. The insight that the governor is spinning too fast or too slowly also needs very little knowledge of the full steam engine.

Fundamentally, both monitoring and observability are used to improve the reliability and performance of the system in question.

Now that we have introduced the high-level concepts, let's explore a practical example outside of the world of software services.

Case study – A ship passing through the Panama Canal

Let's imagine a ship traversing the Agua Clara locks on the Panama Canal. This can be illustrated using the following figure:

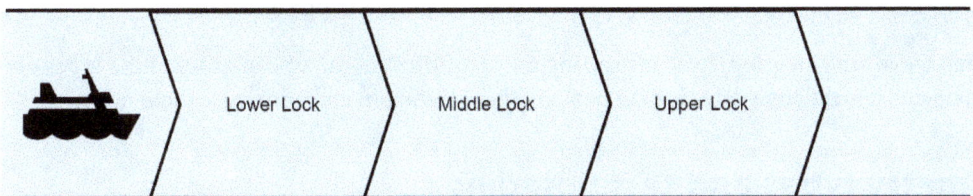

Figure 1.2 – The Agua Clara locks on the Panama Canal

There are a few aspects of these locks that we might want to monitor:

- The successful opening and closing of each gate
- The water level inside each lock
- How long it takes for a ship to traverse the locks

Monitoring these aspects may highlight situations that we need to be alerted about:

- A gate is stuck open because of a mechanical failure
- The water level is rapidly descending because of a leak
- A ship is taking too long to exit the locks because it is stuck

There may be situations where the data we are monitoring are within acceptable limits, but we can still observe a deviation from what is considered *normal*, which should prompt further action:

- A small leak has formed near the top of the lock wall:

  - We would see the water level drop but only when it is above the leak
  - This could prompt maintenance work on the lock wall

- A gate in one lock is opening more slowly because it needs maintenance:

  - We would see the time between opening and closing the gate increase
  - This could prompt maintenance on the lock gate

- Ships take longer to traverse the locks when the wind is coming from a particular direction:

  - We could compare hourly average traversal rates
  - This could prompt work to reduce the impact of wind from one direction

Now that we've seen an example of measuring a real-world system, we can group these types of measurements into different data types to best suit the application. Let's introduce those now.

Telemetry types and technologies

The boring but important part of observability tools is **telemetry** – capturing data that is useful, shipping it from place to place, and producing visualizations, alerts, and reports that offer value to the organization.

Three main types of telemetry are used to build monitoring and observability systems – metrics, logs, and distributed traces. Other telemetry types may be used by some vendors and in particular circumstances. We will touch on these here, but they will be explored in more detail in *Chapters 12* and *13* of this book.

Metrics

Metrics can be thought of as numeric data that is recorded at a point in time and enriched with labels or dimensions to enable analysis. Metrics are frequently generated and are easy to search, making them ideal for determining whether something is wrong or unusual. Let's look at an example of metrics showing temporal changes:

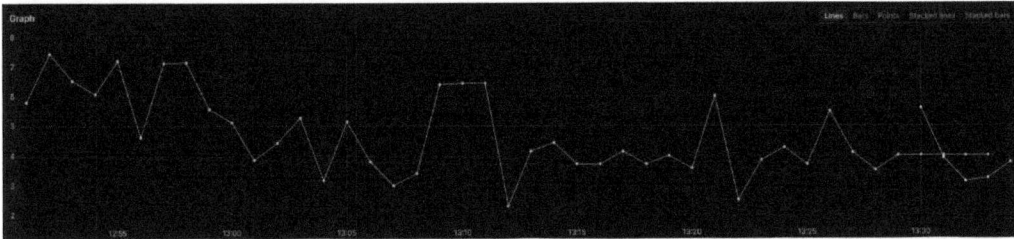

Figure 1.3 – Metrics showing changes over time

Taking our example of the Panama Canal, we could represent the *water level in each lock* as a metric, to be measured at regular intervals. To be able to use the data effectively, we might add some of these labels:

- **The lock name**: Agua Clara
- **The lock chamber**: Lower lock
- **The canal**: Panama Canal

Logs

Logs are considered to be unstructured string data types. They are recorded at a point in time and usually contain a huge amount of information about what is happening. While logs can be structured, there is no guarantee of that structure persisting, because the log producer has control over the structure of the log. Let's look at an example:

```
Jun 26 2016 20:31:01 pc-ac-g1 gate-events no obstructions seen
Jun 26 2016 20:32:01 pc-ac-g1 gate-events starting motors
Jun 26 2016 20:32:30 pc-ac-g1 gate-events motors engaged successfully
Jun 26 2016 20:35:30 pc-ac-g1 gate-events stopping motors
Jun 26 2016 20:35:30 pc-ac-g1 gate-events gate open complete
```

In our example, the various operations involved in opening or closing a lock gate could be represented as logs.

Almost every system produces logs, and they often give very detailed information. This is great for understanding what happened. However, the volume of data presents two problems:

- Searching can be inefficient and slow.

- As the data is in text format, knowing what to search for can be difficult. For example, `error occurred`, `process failed`, and `action did not complete successfully` could all be used to describe a failure, but there are no shared strings to search for.

Let's consider a real log entry from a computer system to see how log data is usually represented:

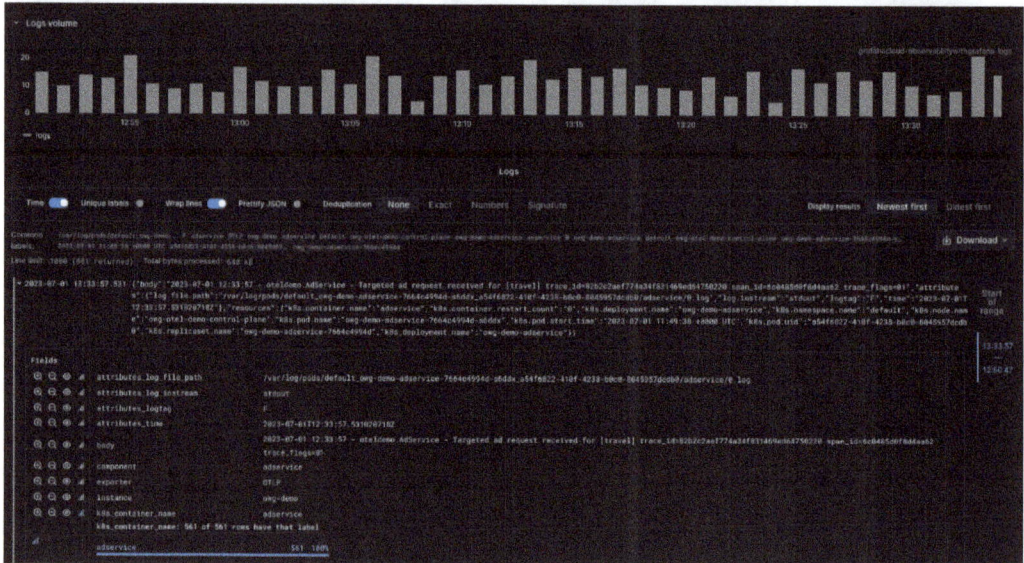

Figure 1.4 – Logs showing discrete events in time

We can clearly see that we have a number of fields that have been extracted from the log entry by the system. These fields detail where the log entry originated from, what time it occurred, and various other items.

Distributed traces

Distributed traces show the end-to-end journey of an action. They are captured from every step that is taken to complete the action. Let's imagine a trace that covers the passage of a ship through the lock system. We will be interested in the time a ship enters and leaves each lock, and we will want to be able to compare different ships using the system. A full passage can be given an identifier, usually called a **trace ID**. Traces are made up of **spans**. In our example, a span would cover the entry and exit for each individual lock. These spans are given a second identifier, called a **span ID**. To tie these two together, each span in a trace references the trace ID for the whole trace. The following screenshot shows an example of how a distributed trace is represented for a computer application:

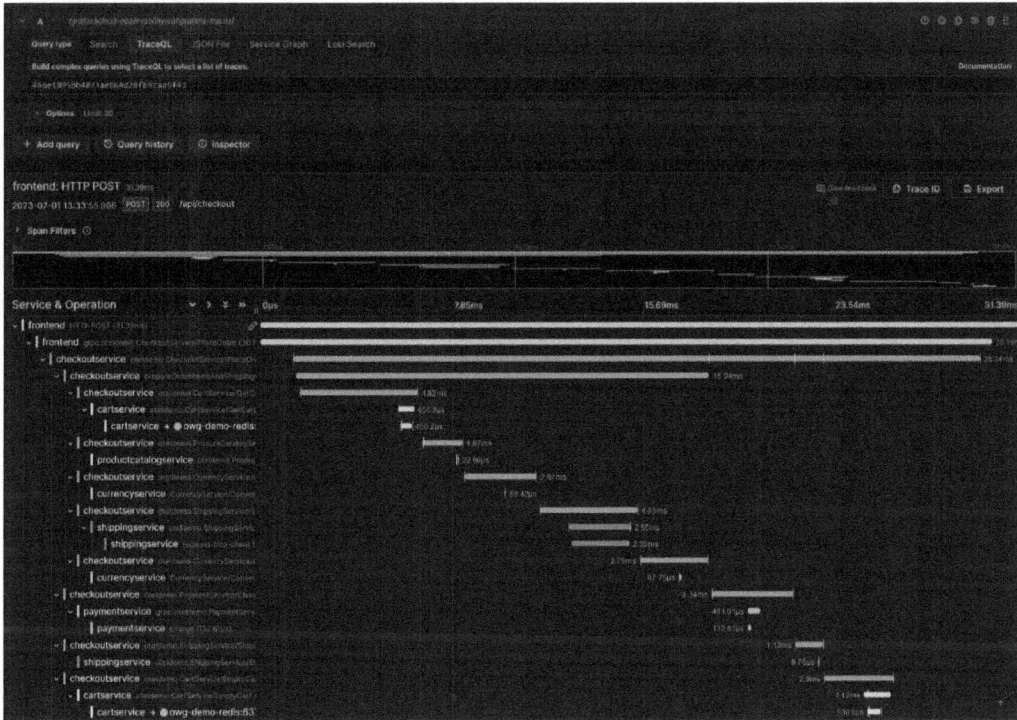

Figure 1.5 – Traces showing the relationship of actions over time

Now that we have introduced metrics, logs, and traces, let's consider a more detailed example of a ship passing through the locks, and how each telemetry type would be produced in this process:

1. Ship enters the first lock:

    - Span ID created

    - Trace ID created

    - Contextual information is added to the span, for example, a ship identification

    - Key events are recorded in the span with time stamps, for example, gates are opened and closed

2. Ship exits the first lock:

    - Span closed and submitted to the recording system

    - Second lock notified of trace ID and span ID

3. Ship enters the second lock:

    - Span ID created

    - Trace ID added to span

    - Contextual information is added to the span

    - Key events recorded in the span with time stamps

4. Ship exits the second lock:

    - Span closed and submitted to the recording system

    - Third lock notified of trace ID and span ID

5. Ship enters the third lock:

    - Repeat *step 3*

6. Ship exits the third lock:

    - Span closed and submitted to the recording system

Now let's look at some other telemetry types.

Other telemetry types

Metrics, logs, and traces are often called the *three pillars* or the *golden triangle* of observability. As we outlined earlier, observability is the ability to understand a system. While metrics, logs, and traces give us a very good ability to understand a system, they are not the only signals we might need, as this depends at what abstraction layer we need to observe the system. For instance, when looking at a very detailed level, we may be very interested in the stack trace of an application's activity at the CPU and RAM level. Conversely, if we are interested in the execution of a CI/CD pipeline, we may just be interested in whether a deployment occurred and nothing more.

Profiling data (stack traces) can give us a very detailed technical view of the system's use of resources such as CPU cycles or memory. With cloud services often charged per hour for these resources, this kind of detailed analysis can easily create cost savings.

Similarly, **events** can be consumed from a platform, such as CI/CD. These can offer a huge amount of insight that can reduce the **Mean Time to Recovery (MTTR)**. Imagine responding to an out-of-hours alert and seeing that a new version of a service was deployed immediately before the issues started occurring. Even better, imagine not having to wake up because the deployment process could check for failures and roll back automatically. Events differ from logs only in that an event represents a whole action. In our earlier example in the *Logs* section, we created five logs, but all of these referred to stages of the same event (opening the lock gate). As a relatively generic term, *event* gets used with other meanings.

Now that we've introduced the fundamental concepts of the technology, let's talk about the customers who will use observability data.

Introducing the user personas of observers

Observability deals with understanding a system, identifying whether something is wrong with that system, and understanding why it is wrong. But what do we mean by *understanding a system*? The simple answer would be *knowing the state of a single application or infrastructure component.*

In this section, we will introduce the user personas that we will use throughout this book. These personas will help to distinguish the different types of questions that people use observability systems to answer.

Let's take a quick look at the user personas that will be used throughout the book as examples, and their roles:

	Name and role	Description
	Diego Developer	Frontend, backend, full stack, and so on
	Ophelia Operator	SRE, DevOps, DevSecOps, customer success, and so on

	Steven Service	Service manager and other tasks
	Pelé Product	Product manager, product owner, and so on
	Masha Manager	Manager, senior leadership, and so on

Table 1.1 – User persona introductions

Now let's look at each of these users in greater detail.

Diego Developer

Diego Developer works on many types of systems, from frontend applications that customers directly interact with, to backend systems that let his organization store data in ways that delight its customers. You might even find him working on platforms that other developers use to get their applications integrated, built, delivered, and deployed safely and speedily.

Goals

He writes great software that is well tested and addresses customers' actual needs.

Interactions

When he is not writing code, he works with *Ophelia Operator* to address any questions and issues that occur.

Pelé Product works in his team and provides insight into the customer's needs. They work together closely, taking those needs and turning them into detailed plans on how to deliver software that addresses them.

Steven Service is keen to ensure that the changes Diego makes are not impacting customer commitments. He's also the one who wakes Diego up if there is an incident that needs attention. The data provided to *Masha Manager* gives her a breakdown of costs. When Diego is working on developer platforms, he also collects data that helps her get investment from the business into teams that are not performing as expected.

Needs

Diego really needs easy-to-use libraries for the languages he uses to instrument the code he produces. He does not have time to become an expert. He wants to be able to add a few lines of code and get results quickly.

Having a clear standard for acceptable performance measures makes it easy for him to get the right results.

Pain points

When Diego's systems produce too much data, he finds it difficult to sort signal from noise. He also gets frustrated having to change his code because of an upstream decision to change tooling.

Ophelia Operator

Ophelia Operator works in an operations-focused environment. You might find her in a customer-facing role or as part of a development team as a DevOps engineer. She could be part of a group dedicated to the reliability of an organization's systems, or she could be working in security or finance to ensure the business runs securely and smoothly.

Goals

Ophelia wants to make sure a product is functioning as expected. She also likes it when she is not woken up early in the morning by an incident.

Interactions

Ophelia will work a lot with *Diego Developer*; sometimes it's escalating customer tickets when she doesn't have the data available to understand the problem; at other times it's developing runbooks to keep the systems running. Sometimes she will need to give Diego clear information on acceptable performance measures so that her team can make sure systems perform well for customers.

Steven Service works closely with Ophelia. They work together to ensure there are not many incidents, and that they are quickly resolved. Steven makes sure that business data on changes and incidents is tracked, and tweaks processes when things aren't working.

Pelé Product likes to have data showing the problematic areas of his products.

Needs

Good data is necessary to do the job effectively. Being able to see that a customer has encountered an error can make the difference between resolving a problem straight away or having them wait maybe weeks for a response.

During an incident seeing that a new version of a service was deployed at the time a problem started can change an hours-long incident into a brief blip, and keep customers happy.

Pain points

Getting continuous alerts but not being empowered to fix the underlying issue is a big problem. Ophelia has seen colleagues burn out, and it makes her want to leave the organization when this happens.

Steven Service

Steven Service works in service delivery. He is interested in making sure the organization's services are delivered smoothly. Jumping in on critical incidents and coordinating actions to get them resolved as quickly as possible is part of the job. So is ensuring that changes are made using processes that help others do it as safely as possible. Steven also works with third parties who provide services that are critical to the running of the organization.

Goals

He wants services to run as smoothly as possible so that the organization can spend more time focused on customers.

Interactions

Diego Developer and *Ophelia Operator* work a lot with the change management processes created by *Steven* and the support processes he manages. Having accurate data to hand during change management really helps to make the process as smooth as possible.

Steven works very closely with *Masha Manager* to make sure she has access to data showing where processes are working smoothly and where they need to spend time improving them.

Needs

He needs to be able to compare the delivery of different products and provide that data to *Masha* and the business.

During incidents, he needs to be able to get the right people on the call as quickly as possible and keep a record of what happened for the incident post-mortem.

Pain points

Being able to identify the right person to get on a call during an incident is a common problem he faces. Seeing incidents drag on while different systems are compared and who can fix the problem is argued about is also a big concern to him.

Pelé Product

Pelé Product works in the product team. You'll find him working with customers to understand their needs, keeping product roadmaps in order, and communicating requirements back to developers such as *Diego Developer* so they can build them. You might also find him understanding and shaping the product backlog for the internal platforms used by developers in the organization.

Goal

Pelé wants to understand customers, give them products that delight them, and keep them coming back.

Interactions

He spends a lot of time working with *Diego* when they can look at the same information to really understand what customers are doing and how they can help them do it better.

Ophelia Operator and *Steven Service* help *Pelé* keep products on track. If too many incidents occur, they ask everyone to refocus on getting stability right. There is no point in providing customers with lots of features on a system that they can't trust.

Pelé works closely with *Masha Manager* to ensure the organization has the right skills in the teams that build products. The business depends on her leadership to make sure that these developers have the best tools to help them get their code live in front of customers where it can be used.

Needs

Pelé needs to be able to understand customers' pain points even when they do not articulate them clearly during user research.

He needs data that gives him a common language with *Diego* and *Ophelia*. Sometimes they can get too focused on specific numbers such as shaving off a couple of milliseconds from a request, when improving a poor workflow would improve the customer experience more significantly.

Pain points

Pelé hates not being able to see at a high level what customers are doing. Understanding which bits of an application have the most usage, and which bits are not used at all, lets him know where to focus time and resources.

While customers never tell him they want stability, if it's not there they will lose trust very quickly and start to look at alternatives.

Masha Manager

Masha works in management. You might find her leading a team and working closely with them daily. She also represents middle management, setting strategy and making tactical choices, and she is involved, to some extent, in senior leadership. Much of her role involves managing budgets and people. If something can make that process easier, then she is usually interested in hearing about it. What *Masha* does not want to do is waste the organization's money, because that can directly impact jobs.

Goals

Her primary goals are to keep the organization running smoothly and ensure the budget is balanced.

Interactions

As a leader, *Masha* needs accurate data and needs to be able to trust the teams who provide that data. The data could be the end-to-end cycle time of feature concept to delivery from *Pelé Product*, the lead time for changes from *Diego Developer*, or even the MTTR from *Steven Service*. Having that data helps her to understand where focus and resources can have the biggest impact.

Masha works regularly with the financial operations staff and needs to make sure they have accurate information on the organization's expenditure and the value that expenditure provides.

Needs

She needs good data in a place where she can view it and make good decisions. This usually means she consumes information from a business intelligence system. To use such tools effectively, she needs to be clear on what the organization's goals are, so that the correct data can be collected to help her understand how her teams are tracking to that goal.

She also needs to know that the teams she is responsible for have the correct data and tools to excel in their given areas.

Pain points

High failure rates and long recovery time usually result in her having to speak with customers to apologize. *Masha* really hates these calls!

Poor visibility of cloud systems is a particular concern. *Masha* has too many horror stories of huge overspending caused by a lack of monitoring; she would rather spend that budget on something more useful.

You now know about the customers who use observability data, and the types of data you will be using to meet their needs. As the main focus of this book is on Grafana as the underlying technology, let's now introduce the tools that make up the Grafana stack.

Introducing the Grafana stack

Grafana was born in 2013 when a developer was looking for a new user interface to display metrics from Graphite. Initially forked from Kibana, the Grafana project was developed to make it easy to build quick, interactive dashboards that were valuable to organizations. In 2014, **Grafana Labs** was formed with the core value of building a sustainable business with a strong commitment to open source projects. From that foundation, Grafana has grown into a strong company supporting more than 1 million active installations. Grafana Labs is a huge contributor to open source projects, from their own tools to widely adopted technologies such as Prometheus, and recent initiatives with a lot of traction such as OpenTelemetry.

Grafana offers many tools, which we've grouped into the following categories:

- The core Grafana stack: **LGTM** and the **Grafana Agent**
- Grafana enterprise plugins
- Incident response tools
- Other Grafana tools

Let's explore these tools in the following sections.

The core Grafana stack

The core Grafana stack consists of Mimir, Loki, Tempo, and Grafana; the acronym LGTM is often used to refer to this tech stack.

Mimir

Mimir is a **Time Series Database** (**TSDB**) for the storage of metric data. It uses low-cost object storage such as S3, GCS, or Azure Blob Storage. First announced for general availability in March 2022, Mimir is the newest of the four products we'll discuss here, although it's worth highlighting that Mimir initially forked from another project, Cortex, which was started in 2016. Parts of Cortex also form the core of Loki and Tempo.

Mimir is a fully Prometheus-compatible solution that addresses the common scalability problems encountered with storing and searching huge quantities of metric data. In 2021 Mimir was load tested to 1 billion active time series. An active time series is a metric with a value and unique labels that has reported a sample in the last 20 minutes. We will explore Mimir and Prometheus in much greater detail in *Chapter 5*.

Loki

Loki is a set of components that offer a full feature logging stack. Loki uses lower-cost object storage such as S3 or GCS, and only indexes label metadata. Loki entered general availability in November 2019.

Log aggregation tools typically use two data structures to store log data. An index that contains references to the location of the raw data paired with searchable metadata, and the raw data itself stored in a compressed form. Loki differs from a lot of other log aggregation tools by keeping the index data relatively small and scaling the search functionality by using horizontal scaling of the querying component. The process of selecting the best index fields is one we will cover in *Chapter 4*.

Tempo

Tempo is a storage backend for high-scale distributed trace telemetry, with the aim of sampling 100% of the read path. Like Loki and Mimir, it leverages lower-cost object storage such as S3, GCS, or Azure Blob Storage. Tempo went into general availability in June 2021.

When Tempo released 1.0, it was tested at a sustained ingestion of >2 million spans per second (about 350 MB per second). Tempo also offers the ability to generate metrics from spans as they are ingested; these metrics can be written to any backend that supports Prometheus remote write. Tempo is explored in detail in *Chapter 6*.

Grafana

Grafana has been a staple for fantastic visualization of data since 2014. It has targeted the ability to connect to a huge variety of data sources from TSDBs to relational databases and even other observability tools. Grafana has over 150 data source plugins available. Grafana has a huge community using it for many different purposes. This community supports over 6,000 dashboards, which means there is a starting place for most available technologies with minimal time to value.

Grafana Agent

Collecting telemetry from many places is one of the fundamental aspects of observability. **Grafana Agent** is a collection of tools for collecting logs, metrics, and traces. There are many other collection tools that Grafana integrates well with. Different collection tools offer different advantages and disadvantages, which is not a topic we will explore in this book. We will highlight other tools in the space later in this chapter and in *Chapter 2* to give you a starting point for learning more about this topic. We will also briefly discuss architecting a collection infrastructure in *Chapter 11*.

The Grafana stack is a fantastic group of open source software for observability. The commitment of Grafana Labs to open source is supported by great enterprise plugins. Let's explore them now.

Grafana Enterprise plugins

As part of their Cloud Pro, Cloud Advanced, and Enterprise license offerings, Grafana offers Enterprise plugins. These are part of any paid subscription to Grafana.

The Enterprise data source plugins allow organizations to read data from many other storage tools they may use, from software development tools such as **GitLab** and **Azure DevOps** to business intelligence tools such as **Snowflake**, **Databricks**, and **Looker**. Grafana also offers tools to read data from many other observability tools, which enables organizations to build comprehensive operational coverage while offering individual teams a choice of the tools they use.

Alongside the data source plugins, Grafana offers premium tools for logs, metrics, and traces. These include access policies and tokens for log data to secure sensitive information, in-depth health monitoring for the ingest and storage of cloud stacks, and management of tenants.

Grafana incident response and management

Grafana offers three products in the **incident response and management** (IRM) space:

- At the foundation of IRM are **alerting rules**, which can notify via messaging apps, email, or Grafana OnCall

- **Grafana OnCall** offers an on-call schedule management system that centralizes alert grouping and escalation routing

- Finally, **Grafana Incident** offers a chatbot functionality that can set up necessary incident spaces, collect timelines for a post-incident review process, and even manage the incident directly from a messaging service

These tools are covered in more detail in *Chapter 9*. Now let's take a look at some other important Grafana tools.

Other Grafana tools

Grafana Labs continues to be a leader in observability and has acquired several companies in this space to release new products that complement the tools we've already discussed. Let's discuss some of these tools now.

Faro

Grafana Faro is a JavaScript agent that can be added to frontend web applications. The project allows for **real user monitoring** (**RUM**) by collecting telemetry from a browser. By adding RUM into an environment where backend applications and infrastructure are instrumented, observers gain the ability to traverse data from the full application stack. Faro supports the collection of the five core web vitals out of the box, as well as several other signals of interest. Faro entered general availability in November 2022. We cover Faro in more detail in *Chapter 12*.

k6

k6 is a load testing tool that provides both a packaged tool to run in your own infrastructure and a cloud **Software as a Service** (**SaaS**) offering. Load testing, especially as part of a CI/CD pipeline, really enables teams to see how their application will perform under load, and evaluate optimizations and refactoring. Paired with other detailed analysis tools such as Pyroscope, the level of visibility and accessibility to non-technical members of the team can be astounding. The project started back in 2016 and was acquired by Grafana Labs in June 2021. The goal of k6 is to make performance testing easy and repeatable. We'll explore k6 in *Chapter 13*.

Pyroscope

Pyroscope is a recent acquisition of Grafana Labs, joining in March 2023. Pyroscope is a tool that enable teams to engage in the continuous profiling of system resource use by applications (CPU, memory, etc.). Pyroscope advertises that with a minimal overhead of ~2-5% of performance, they can collect samples as frequently as every 10 seconds. **Phlare** is a Grafana Labs project started in 2022, and the two projects have now merged. We discuss Pyroscope in more detail in *Chapter 13*.

Now that you know the different tools available from Grafana Labs, let's look at some alternatives that are available.

Alternatives to the Grafana stack

The monitoring and observability space is packed with different open and closed source solutions such as ps and top going back to the 70s and 80s. We will not attempt to list every tool here; we aim to offer a source of inspiration for people who are curious and want to explore, or who need a quick reference of the available tools (as the authors have on a few occasions).

Data collection

These are agent tools that can be used to collect telemetry from the source:

Tool Name	Telemetry Types
OpenTelemetry Collector	Metrics, logs, traces
FluentBit	Metrics, logs, traces
Vector	Metrics, logs, traces
Vendor-specific agents (See the *Data storage, processing, and visualization* section for an expanded list)	Metrics, logs, traces
Beats family	Metrics, logs
Prometheus	Metrics
Telegraf	Metrics
StatsD	Metrics
Collectd	Metrics
Carbon	Metrics
Syslog-ng	Logs
Rsyslog	Logs
Fluentd	Logs
Flume	Logs
Zipkin Collector	Traces

Table 1.2 – Data collection tools

Data collection is only one piece of the extract transform and load process for observability data. The next section introduces tools to transform and load data.

Data storage, processing, and visualization

We've grouped data processing, storage, and visualization together, as there are often a lot of crossovers among them. There are certain tools that also provide **security monitoring** and are closely related. However, as this topic is outside of the scope of this book, we have chosen to exclude tools that are solely in the security space.

Tool Name	Tool Name	Tool Name
AppDynamics	InfluxDB	Sematext
Aspecto	Instana	Sensu
AWS CloudWatch & CloudTrail	Jaeger	Sentry
Azure Application insights	Kibana	Serverless360
Centreon	Lightstep	SigNoz
ClickHouse	Loggly	SkyWalking
Coralogix	LogicMonitor	Solarwinds
Cortex	Logtail	Sonic
Cyclotron	Logz.io	Splunk
Datadog	Mezmo	Sumo Logic
Dynatrace	Nagios	TelemetryHub
Elastic	NetData	Teletrace
GCP Cloud Operations Suite	New Relic	Thanos
Grafana Labs	OpenSearch	Uptrace
Graphite	OpenTSDB	VictoriaMetrics
Graylog	Prometheus	Zabbix
Honeycomb	Scalyr	Zipkin

Table 1.3 – Data storage processing and visualization tools

With a good understanding of the tools available in this space, let's now look at the ways we can deploy the tools offered by Grafana.

Deploying the Grafana stack

Grafana Labs fully embraces its history as an open source software provider. The LGTM stack, alongside most other components, is open source. There are a few value-added components that are offered as part of an enterprise subscription.

As a SaaS offering, Grafana Labs provides access to storage for Loki, Mimir, and Tempo, alongside Grafana's 100+ integrations for external data sources. As a SaaS customer, you also gain ready access to a huge range of other tools you may use and can present them in a consolidated manner, in a single pane of glass. The SaaS offering allows organizations to leverage a full-featured observability platform without the operational overhead of running the platform and obtaining service level agreements for the operation of the platform.

As well as managing the platform for you, you can run Grafana on your organization's infrastructure. Grafana offers its software packaged in several formats for Linux and Windows deployments, as well as offering containerized versions. Grafana also offers Helm and Tanka configuration wrappers for each of their tools. This book will mainly concentrate on the SaaS offering because it is easy to get started with the free tier. We will explore some areas where a local installation can assist in *Chapters 11* and *14*, which cover architecting and supporting DevOps processes respectively.

Summary

In this chapter, you have been introduced to monitoring and observability, how they are similar, and how they differ. The Agua Clara locks on the Panama Canal acted as a simplified example of the concepts of observability in practice. The key takeaway should be to understand that even when a system produces alerts for significant problems, the same data can be used to observe and investigate other potential problems.

We also talked about the customers who might use observability systems. These customers will be referenced throughout this book when we explore a concept and how to target its implementation.

Finally, we introduced the full Grafana Labs stack, and you should now have a good understanding of the different purposes that each product serves.

In the next chapter, we will introduce the basics of adding instrumentation to applications or infrastructure components for readers whose roles are similar to those of Diego and Ophelia.

2

Instrumenting Applications and Infrastructure

The previous chapter introduced observability, with examples outside of the computing world to give you a generic understanding of the subject. In this chapter, we'll build on those examples by providing a high-level overview of both application and infrastructure instrumentation. We will look at the data created by systems and how that fits into the different telemetry types and common protocols in use. We will also explore widely used libraries for popular programming languages that simplify instrumenting applications. To finish, we will cover more traditional telemetry collection from infrastructure components, operating systems, and network devices. This will give you insight into the components that are still in operation today that run applications and Kubernetes workloads. This chapter is aimed at readers of all technical abilities and no specific technologies are needed. An understanding of observability terminology (for example, logs, metrics, traces, and instrumentation) is helpful. It aims to provide an overview of the technology space and act as a valuable resource that you can quickly reference when you are working with your observability solutions.

In this chapter, we explore the following introductory sections:

- Common log formats
- Metrics protocols and best practices
- Tracing protocols and best practices
- Using libraries to instrument efficiently
- Infrastructure data technologies

Common log formats

Log files are a standard component of computer systems and an essential tool for software developers and operators – in our example, Diego and Ophelia, respectively. Logs support performance and capacity monitoring in infrastructure, bug detection in software, root cause analysis, user behavior tracking, and more. There is no perfect recipe for logs and as such, it does not matter what your logs look like, though following certain guidelines will help your future self when you need to analyze logs. In this section, we will learn about different log formats and how the data can be used. Log formats are the definition of what a log file looks like and should explain how the data can be interpreted.

Log formats usually identify if they are structured or unstructured, the data types used in them, and if any encoding or delimitation is being used. We'll explore structure first and then look at example log formats in more detail in the following sections.

Structured, semi-structured, and unstructured logging

As mentioned previously, it does not matter what your logs look like and they can come in structured, semi-structured, or unstructured formats. However, when designing and building observability solutions, it's important to understand the log formats you are working with. This ensures that you can ingest, parse, and store the data in a way that it can be used effectively. If you familiarized yourself with the personas in *Chapter 1*, you have an awareness of who they will be used by and for what purpose.

Structured logging

Structured logs have a predetermined message format that allows them to be treated as datasets rather than text. The idea of structured logging is to present data with a defined pattern that can be easily understood by humans and efficiently processed by machines. The log entries are often delimited with characters such as a comma, space, or hyphen. Data fields may also be joined using an equals sign or colon for key-value pairs, such as name=Diego or city=Berlin.

Here is an example of a structured log format:

```
{
"timestamp": "2023-04-25T12:15:03.006Z",
"message": "User Diego.Developer has logged in",
"log": {
"level": "info",
"file": "auth.py",
"line": 77
},
"user": {
"name": "diego.developer",
"id": 123
},
```

```
"event": {
"success": true
}
}
```

An additional benefit of structured logging is that you can validate the conformation of the data to a schema with tools such as JSON schema. This opens up the possibility of making version control changes to the schema, which is where logs and event bus technology overlap.

Semi-structured logging

Semi-structured logs aim to bridge the gap between unstructured and structured and, as a result, can be quite complicated. They are designed to be easy for humans to read but also have a schema that makes it possible for machines to process them too. They have complex field and event separators and usually come with a defined pattern to aid with ingesting and parsing. Parsing is usually done using regular expressions or other code.

Unstructured logging

Unstructured logging typically refers to log entries that are presented in a textual format that can easily be read by humans but is difficult for machines to process. They are often color-coded with blank spaces to improve presentation and readability. It is this presentation that creates issues for machines to process the logs. Parsing and splitting the data correctly creates a disassociation between events and their identifying metadata. An unstructured log will require some custom parsing, requiring intimate knowledge of the data and often creating additional work for the engineer (*Ophelia*) when ingesting data. This also creates technical liability; the dependency on the log remaining the same restricts developers from changing logs or runs the risk of parsing and reporting on unstructured logs prone to breaking.

To aid the ability of machines to process unstructured logs, encapsulation prevents entries such as stack traces from splitting at an inappropriate location.

The following is an example of a multiline log, with a naive encapsulation that looks for line breaks; this will appear in logging systems as four distinct events:

```
2023-04-25 12:15:03,006 INFO [SVR042] UserMembershipsIterable Found 4
children for 4 groups in 3 ms
Begin Transaction update record.
Process started.
Process completed.
```

With encapsulation based on the timestamp at the start of the event, this will be stored correctly for searching.

In the following section, we will explore common log formats found in today's systems.

Sample log formats

Many log formats have been used in computer systems. All of these formats have a common goal of presenting a standard structure or set of fields for recording important information about the activity of a computer system. The following table aims to provide easy reference for some of the more notable ones:

Format	Overview
Common Event Format (CEF)	CEF is an open logging and auditing format from ArcSight that aims to provide a simple interface to record security-related events.
NCSA **Common Log Format (CLF)**	The NCSA CLF is historically used on web servers to record information about requests made to the server. This format has been extended by the CLF to include additional information about the browser (user-agent) and the referer.
W3C Extended Log File Format	W3C Extended Log File Format is a log format commonly used by Windows Internet Information Services servers (web servers).
Windows Event Log	Windows Event Log is the standard log format used by the Windows operating system. These logs record events that occur on the system and are categorized System, Application, Security, Setup, and Forwarded events.
JavaScript Object Notation (JSON)	JSON is an open standard file format that is very useful for easily parsing structured log events.
Syslog	Syslog is a standard that's used across many hardware devices such as networking, compute, and storage, and is used by the Linux kernel for logging.
Logfmt	Logfmt does not have a defined standard but is a widely used form of human-readable structured logging.

Table 2.1 – Log format overview

Let's look at these formats in greater detail.

CEF

Developed by ArcSight to fulfill the **Security Information and Event Management (SIEM)** use case, the CEF is a structured text-based log format. Using UTF-8 encoding, the format contains a prefix, a CEF header, and a body containing additional enrichment data.

The following table shows the log sections of the CEF format:

Log Section	Description
Prefix	It combines the event timestamp and source hostname.
CEF header	It combines the following pieces of metadata: • Software version • Vendor name • Product name • Product version • Product event class identification code • Event name • Event severity
Body	It contains a list of key-value pairs

Table 2.2 – CEF format

Here is an example CEF log event:

```
CEF:0|Security Provider|Security Product|Version|123|User
Authenticated|3|src=10.51.113.149 suser=diego target=diego msg=User
authenticated from 1001:db7::5
```

NCSA CLF

As one of the oldest log formats used by web servers, the NCSA CLF has for a long time been the most common and well-known log formats. It has a fixed format text-based structure and therefore cannot be customized at all.

Here is the NCSA CLF field list:

- Remote host address

- Remote log name

- Username

- Timestamp

- Request and protocol version

- HTTP status code

- Bytes sent

Where data is missing from the log, a hyphen acts as a placeholder. Unsupported characters are replaced with the + symbol.

Here is an example NCSA CLF log:

```
127.0.0.1 user-identifier diego [25/Apr/2023:12:15:03 -0000] "GET /
apache_pb.gif HTTP/1.1" 200 2326
```

W3C Extended Log File Format

The Microsoft Internet Information Server log format known as W3C is a structured yet configurable format. Full control over the included fields ensures log files contain the most relevant data. Identification of the information or direction of flow is denoted using a string prefix: server (*S*), client (*C*), server to client (*SC*), and client to server (*CS*).

Here is the W3C Extended Log File Format field list:

- Timestamp

- Client IP

- Server IP

- URI-stem

- HTTP status code

- Bytes sent

- Bytes received

- Time taken

- Version

Here is an example W3C log:

```
#Software: Internet Information Services 10.0
#Version: 1.0
#Date: 2023-04-25 12:15:03
#Fields: time c-ip cs-method cs-uri-stem sc-status cs-version
12:15:03 10.51.113.149 GET /home.htm 200 HTTP/1.0
```

Microsoft Windows Event Log

The Microsoft Windows operating system comes with a built-in complex structured logging system that captures data related to specific events on the operating system. There are four common Windows event log categories – system, application, security, and setup – and an additional special category for forwarded events.

Each event log is also one of five different types: information, warning, error, success audit, and failure audit. Windows Event Log is one of the most verbose log formats in use. It usually includes details such as timestamp, event ID, username, hostname, message, and category, making it invaluable in diagnosing problems. Windows event IDs are documented and searchable, so you can easily get detailed information regarding the log event; they are grouped into categories, narrowing down the area where the event occurred, which makes debugging very accurate.

Here is a trimmed example of Microsoft Windows Event Log:

```
An account was successfully logged on.
Subject:
Security ID: SYSTEM
Account Name: DESKTOP-TMC369$
Account Domain: WORKGROUP
Logon ID: 0xE37
Logon Information:
New Logon:
Security ID: AD\DiegoDeveloper
Account Name: diego.developer@themelt.cafe
Account Domain: AD
Logon ID: 0xEC4093F
Network Information:
Workstation Name: DESKTOP-TMC369
```

JSON

As one of the newer yet most commonly used log formats today, JSON is a structured format constructed from multiple key-value pairs. Using JSON, data can be nested into different layers while keeping the format easy to read. Additionally, different data types can be represented, such as string, number, Boolean, null, object, and array.

Here is an example JSON log file:

```
{
"timestamp": "2023-04-25T12:15:03.006Z",
"message": "User Diego.Developer has logged in",
"log": {
"level": "info",
```

```
"file": "auth.py",
"line": 77
},
"user": {
"name": "diego.developer",
"id": 123
},
"event": {
"success": true
}
}
```

Syslog

The go-to log format for many years and still widely used, Syslog is a defined standard for creating and transmitting logs. The **Syslog transport protocol** specifies how log transmission takes place, as well as the data format. The default network ports for the protocol are 514 and 6514, with the latter being used for encryption.

The Syslog message format combines a standardized header and message holding the body of the log.

Here is an example Syslog log:

```
Apr 25 12:15:03 server1 sshd[41458] : Failed password for  diego from
10.51.113.149 port 22 ssh2
```

Logfmt

Logfmt is a widely used log format that fits as human readable and structured so that computers and people can both read it. A Logfmt-formatted log line consists of any number of key-value pairs that can be easily parsed. As there are no standards, it is easy to extend and perfect for developers to simply add more key-value pairs to the output.

Here is an example Logfmt log:

```
level=info method=GET path=/ host=myserver.me fwd="10.51.113.149"
service=4ms status=200
```

Exploring metric types and best practices

Metrics, along with logs, are an essential tool for software developers (*Diego*) and operators (*Ophelia*), providing them with indicators regarding the state of applications and systems. Resource usage data is great for monitoring a metric that captures numerical data over time. There are many different types of resources but some good examples would be CPU or RAM usage, the number of messages in a queue, and the number of received HTTP requests. Metrics are frequently generated and easily enriched with labels, attributes, or dimensions, making them efficient to search and ideal in determining if something is wrong, or different from usual.

A metric commonly has the following fields:

- **Name**: This uniquely identifies the metric
- **Data point value(s)**: The data that's stored varies by metric type
- **Dimensions**: Additional enrichment labels or attributes that support analysis

Metrics capture the behavior of the data they represent. CPU usage would go up and down between 0% and 100% usage, whereas the number of received HTTP requests could increase indefinitely. In the following section, we will look at metric types, which allow us to capture the behavior of the metric being collected.

Metric types

Metrics vary in characteristics and structure. There are four common types of metrics, from simple single values to more complex values:

- **Counter**: This metric represents the last increment value. This could be the incremental change from the last recording or the total increment since the recording started.

  Here are some examples of this metric:

  - The number of requests served
  - Tasks completed
  - Errors reported

  How the value is reset to zero depends on the protocol used to collect them, so it is important to factor this in for your use case. The StatsD implementation resets the counter every time the value is flushed, and Prometheus resets the counter when the application process restarts.

- **Gauge**: A gauge metric is a snapshot of state and can be used to take a measure of something reporting continuously. As such, it is usually made more useful by aggregating with sum, average, minimum, or maximum over a certain period.

  Here are some examples of this metric:

  - Temperature
  - Items in queue
  - Disk space used
  - Number of concurrent requests

Like counter, the definitions for gauge vary in implementation, so be sure to verify how the protocol you select will report gauge metrics.

- **Histogram**: A histogram metric represents the statistical distribution of a set of values returning, for example, min, max sum, and count data points. These are calculated by the agent, reported in a single time interval, and often counted in configurable buckets. They return the raw values; this differs from a summary, which returns the percentile values. Here are some examples:

  - Request durations

  - Response sizes

  In Prometheus, a histogram is made up of a count of the total measurements (`_count`), a sum of all the values of the measurements (`_sum`), and several buckets that have a count of events with a measure less than or equal (`le`) to a defined value.

  Definitions can vary in implementation – for example, Prometheus has a `histogram_quantile` function that can be used to calculate percentiles from histogram metrics.

- **Summary**: Similar to a histogram, a summary samples a set of values. While it provides sum and count, it also calculates percentiles over a sliding time window. These are usually a good option if you need to calculate accurate percentiles but cannot be sure what the range of the values will be. Some examples of this metric are as follows:

  - Request durations

  - Response sizes

  In Prometheus, a summary, like a histogram, is made up of the `_count` and `_sum` metrics and several groupings. Unlike a histogram, these groupings are a quantile, and the value represents the value of that quantile at the point in time for the measurement. For example, a quantile of 0.99 and a value of 3.2148 would indicate that 99% of the sampled data was smaller than 3.2148.

Again, definitions can vary in implementation, so work out what your goals are from your metrics to ensure the capabilities are supported by your choice of protocol. It's useful to note that, in Prometheus, summary metrics have a significant drawback in modern systems as they cannot be aggregated over multiple sources.

There are some distinct differences between these metric types, as we will discuss in the following section.

Comparing metric types

The following table describes each type in general terms. When querying them, this provides a useful reference when approaching metric adoption:

Consideration	Counter	Gauge	Histogram	Summary
Structure	Simple	Simple	Complex	Complex
Can increase and decrease	No	Yes	No	No
Is an approximation	No	No	Yes	Yes
Can calculate percentiles	No	No	Yes	Yes
Can use a rate function	Yes	No	No	No
Can be queried with the `prometheus histogram_ quantile` function	No	No	Yes	No
Can be aggregated across multiple series	Yes	Yes	Yes	No

Table 2.3 – Comparison of metric types

The following table provides a few reference examples of the type and values expected:

Metric Type	Data Field	Value
Counter	Last increment	15
Gauge	Last value	25.4
Histogram	Min	0
	Max	100
	Count	10
	Interval	20
	0-20	1
	20-40	2
	40-60	4
	60-80	2
	80-100	1
Summary	Min	1.2ms
	Max	4.23ms
	Count	10

	Sum	
	Percentiles/Quantiles	
	P90	2.98ms
	P95	3.76ms
	P99	4.23ms

Table 2.4 – Metric type example data

Now that we've looked at the different types of metrics, let's look at the different technologies used to transmit metrics.

Metric protocols

Metric protocols are collections of tools and libraries for instrumenting applications, data formats to transmit, clients to collect data, and often storage and visualization tools. Some common protocols that are in use today are described in the following table:

Metric Protocol	Features
StatsD	It supports the following: • Counters • Gauges • Timers • Histograms • Meters
DogStatsD	DogStatsD implements the StatsD protocol and adds a few Datadog-specific extensions: • Histogram metric type • Service checks • Events • Tagging

	It supports the following:
OpenTelemetry Protocol (OTLP)	<ul><li>Counters</li><li>Gauges</li><li>Histograms</li><li>Summaries (legacy support)</li></ul>
Prometheus	It supports the following:<ul><li>Counters</li><li>Gauges</li><li>Cumulative histograms</li><li>Summaries</li></ul>

Table 2.5 – Common metric protocols and their features

Metrics are very powerful, but some pitfalls can catch people out. Some of these can lead to expensive mistakes. To avoid these pitfalls, let's discuss some best practices.

Best practices for implementing metrics

Introducing metrics into your services is a very good way to gain a huge amount of visibility on how they behave in real situations. The following best practices are from our experience with metrics and will help you manage scope creep, cost, and linking metrics up with traces:

- **Set your objectives**: Work out what your objectives are from your metrics. We have already spoken about the variation in implementation between metric protocols – this can have a big impact if you are expecting to use a metric in a certain way and haven't factored in nuances.

  This will also help you define **service-level indicators** (**SLIs**) and **service-level objectives** (**SLOs**), which will be useful in *Chapter 9, Managing Incidents Using Alerts*.

- **Manage cardinality**: Cardinality is generally defined as the number of unique elements in a set. High cardinality may provide richer, more useful data, but at the cost of monitoring performance impacts or increased storage costs. For example, if you dimension your metrics by server name, the sample could be small, maybe a few hundred metrics. If we compare this to dimensioning by user, which could be in the millions, the increase in the number of metrics produced is exponential. This increase has a direct impact on load and storage.

  Take time to understand the capabilities of the observability backend – things such as the billing framework, limitations, storage, and performance.

- **Add context**: The ability to correlate (establish a common identifier) metrics with traces has been introduced to Grafana and Open Telemetry recently with exemplars. They enable quick visualization and linking between a metric data point and a specific trace span, thus giving improved context and detail to your data.

As we just discussed, metrics capture numerical data from a single service; however, the systems that operate today may consist of multiple services. Distributed tracing is a way to gain visibility of the communications between services. Let's take a look at tracing protocols and some best practices regarding them.

Tracing protocols and best practices

Tracing, or as it is more commonly referred to, **distributed tracing**, tracks application requests as they are made between services of a system. It allows you to follow a single request through an entire system or look at the aggregate data over requests to better understand distributed behavior.

This capability provides software developers (*Diego*), operators (*Ophelia*), and service managers (*Steven*) with valuable tools that enable an understanding of the flow of logic that is essential for troubleshooting. Instrumenting your code by adding traces helps you easily pinpoint almost any issue or at least have a clear indicator of where the problem could be. **Distributed tracing** uses the concepts of spans and traces to capture this data. Let's examine these in more detail.

Spans and traces

The **trace record** is the parent object that represents the data flow or execution path through the system being observed. Each **trace** will contain one or more **span** records that represent the logical operations. This relationship between traces and spans is illustrated in the following figure, in what can be thought of as a directed acyclic graph of spans:

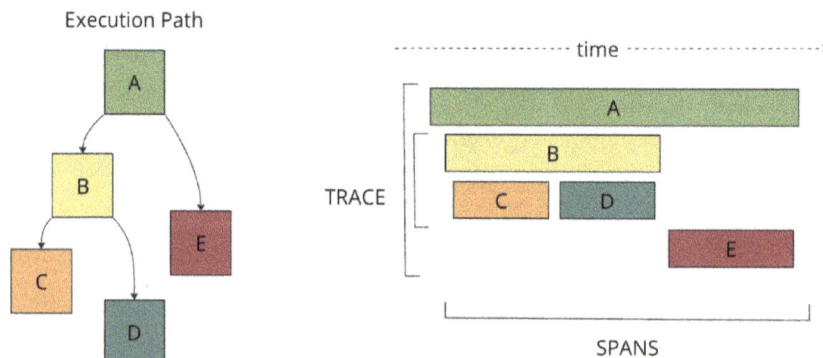

Figure 2.1 – Traces and spans

A trace is pieced together from multiple spans and would usually report the following information:

- **Identifier**: Uniquely identifies the trace
- **Name**: Describes the overall work being recorded
- **Timing details**: Provides the start and end timestamps for the complete trace

A span commonly has the following fields:

- **Trace identifier**: Establishes the trace relationship
- **Identifier**: Uniquely identifies the span
- **Parent span identifier**: Establishes a parent relationship
- **Name**: Describes the work being recorded
- **Timing details**: Provides the start and end timestamps

A trace identifier will be automatically generated if one has not been received by the calling operation; each application will pass the trace ID along to the next.

The start and end timestamps for the operation help identify which stages are taking the most time. You can drill down to identify dependencies on other services and how they contribute to the overall trace timings.

Spans can often have additional fields that are specific to the protocol implemented. Investigating the options against your use case will help provide the right diagnostics for your system.

Tracing protocols

As with all technology, standards have taken a while to be formalized for tracing, and a few protocols have been implemented. Some common protocols that are in use today are described in the following table:

Protocol Name	Features
OTLP	It supports the following: • Additional fields • Span attributes (metadata about the operation) • Context propagation • Span events (meaningful point-in-time annotation) • Span links (imply a causal relationship between spans) • Span kind (additional details supporting the assembly of a trace)

Zipkin	It supports the following: • Additional fields • Span tags (metadata about the operation) • Context propagation • Span annotations (such as OTLP events and meaningful point-in-time annotation) • Span kind (additional details supporting the assembly of a trace)
Jaeger	It supports two formats – Jaeger Thrift and Jaeger Proto – with similar characteristics. Jaeger Proto has been discontinued in favor of OTLP. It supports the following: • Additional fields • Span tags (metadata about the operation) • Context propagation (Thrift only; Proto does not support this) • Span logs (meaningful point-in-time annotation) • Span references (imply a causal relationship between spans) • Span kind (similar to OTLP, this is stored as a special type of span tag)

Table 2.6 – Distributed tracing protocols and features

Implementing distributed tracing can be a daunting task, so let's discuss some best practices that will help you avoid common mistakes and issues.

Best practices for setting up distributed tracing

So far, we have described how traces will help you with problem resolution. However, when producing traces, it's worth considering the additional system visibility against cost and performance impacts. Let's discuss some of the best practices that should be considered when implementing tracing on any application or system.

Performance

The process of generating trace information can potentially incur a performance overhead at the application level. Mix this with the reduced level of control with auto-instrumentation and the problem can increase.

Here are some of the possible impacts to consider:

- Increased latency
- Memory overhead
- Slower startup time

Some of the more recent observability agents have addressed a lot of the issues with configurable options. For example, the OpenTelemetry Collector offers a sampling configuration that will submit 0% to 100% of spans to the collection tool. This sampling implementation will also notify any downstream services that the parent sampled its span so that the full trace will be collected.

Cost

Increased network and storage costs can become a factor and need factoring in as a limitation when designing your observability solution. However, this does depend on your observability backend and if you are doing additional processing or filtering when the data is being transmitted.

The mitigation practices are as follows:

- **Sampling**: Only sends a percentage of traces
- **Filtering**: Restricts which traces are transmitted and stored
- **Retention**: Sets optimal data storage durations

Accuracy

To ensure one of the major benefits of tracing is implemented, it is important to ensure context propagation is working correctly. Without the relationships being established between the operations, spans will be broken across multiple traces. Validating and solving this problem will increase the usability and adoption of tracing for fast issue resolution.

With most code, libraries are used so that developers can focus on writing code that provides value to the organization. The modern libraries that are available will help you instrument quickly so that you can start using the data collected from your application. We'll explore this next.

Using libraries to instrument efficiently

Instrumenting your application code to emit the telemetry of logs, metrics, and traces can be complex, time-consuming, and difficult to maintain. There are two main approaches to solving this problem – automatic instrumentation and manual instrumentation – with a wide selection of SDKs and libraries available to support them. Here is a brief overview of them:

- **Automatic instrumentation**: Automatic instrumentation is the simplest to implement but can lack the level of control that's often required when building an observability platform. In a very short space of time, it will provide visibility into your application and help you start answering your observability questions. Without careful configuration and design, this will lead to other problems such as performance and cost issues, and, in the worst case, render the observability platform useless.

The approach varies depending on the programming language; for example, code manipulation (during compilation or at runtime) is often used with Java, whereas monkey patching (updating behavior dynamically at runtime) is often used with Python and JavaScript.

- **Manual instrumentation**: Manual instrumentation can be quite complex, depending on the systems being instrumented. It requires an intimate knowledge of the application code, with the benefit of allowing you to specify exactly what telemetry you want. Additionally, you need to understand the observability API you are working with. Though SDKs and libraries have simplified this, a lot of work must be done to understand the implementation.

If you are interested in further reading about application instrumentation, there is an excellent section dedicated to the subject in Alex Boten's book *Cloud-Native Observability with OpenTelemetry*, by Packt Publishing.

Now that we've seen how various libraries approach instrumentation, let's look at some of the common libraries that are used in different languages.

Popular libraries for different programming languages

There have been many telemetry solutions, SDKs, and libraries over the years; however, in more recent history, there has been a concerted effort to align on supporting the OpenTelemetry standard. With its goal to provide a set of standardized vendor-agnostic SDKs, APIs, and tools for ingesting, transforming, and transporting data to an observability backend platform, there are obvious benefits. We will look at the OpenTelemetry libraries in this section to focus on where the most enhancements are currently. However, investigating what is appropriate for your use case is important. One drawback of this concerted development effort is that it creates a fast-changing landscape, so you have to pay attention to release stability and monitor for changes and improvements.

Here are some of the available instrumentation libraries:

Language	SDKs and Libraries	Notes
JavaScript	OpenTelemetry JavaScript SDK	Multiple resources and examples are available that cover Node.js and browser implementations.
JavaScript	OpenTelemetry JavaScript Contrib	An additional repository for OpenTelemetry JavaScript contributions that are not part of the core repository and core distribution of the API and the SDK.

Python	OpenTelemetry Python SDK	At the time of writing, both traces and metrics are stable, with logs in an experimental state.
Python	OpenTelemetry Python Contrib	An additional repository for OpenTelemetry Python contributions. At the time of writing, Contrib libraries are in beta and active development.
Java	OpenTelemetry Java SDK	There is a long list of supported libraries and frameworks with good documentation to get you started.
Java	Spring Boot/Micrometer	As of Spring Boot 3, the default exporter for Micrometer is OTLP.

Table 2.7 – Common libraries and SDKs for telemetry

Applications are only one part of the computer systems we work with today. Our infrastructure components, such as switches, servers, Kubernetes clusters, and more, are just as important to observe. We'll discuss how we can do this in the next section.

Infrastructure data technologies

So far in this chapter, we have focused on implementations that work well for cloud technologies and containerized platforms. Underneath all of the abstraction are physical components, the servers running the workloads, the network and security devices handling communications, and the power and cooling components that keep things running. These have not dramatically changed over time and neither has the telemetry reported by the logs and metrics. Let's take a look at the common infrastructure components and standards used in this area.

Common infrastructure components

Infrastructure can largely be categorized into some broad categories, as we will discuss in the following sections. The types of data you can collect will differ on the category of the component.

Compute or bare metal

Servers are often referred to as **bare metal** or compute; these are physical devices that are used for computation. Often, these systems would run virtualized operating systems that can collect server telemetry. Usually, you will run an agent on the operating system that scrapes metrics or reads log files and then transports them to a receiver. The data that's obtained from server equipment can not only help in diagnosing and responding to issues but can help predict capacity problems that may arise in the future. Often, these devices can send data outside of any virtual operating system as well.

For instance, here are a few telemetry examples that can indicate if a system is close to capacity in any area:

- System temperature
- CPU utilization percent
- Overall disk space used and remaining
- Memory usage and free memory

Network devices

Network and security devices such as switches and firewalls come with the capability to send monitoring information via SNMP to a receiver. Firewalls can often send Syslog-formatted logs to a receiver. The telemetry provided helps diagnose issues with connectivity – for example, latency and throughput are difficult to investigate without information from the hardware.

Here are some telemetry examples:

- Latency
- Throughput
- Packet loss
- Bandwidth

Power components

The components that provide power or cooling are often built with the capability to emit telemetry over SNMP to a receiver. Some older components will implement the Modbus protocol and expose registers that can be read to obtain metrics. The telemetry reported at this level is simplistic but essential when you are operating your data center. If, for example, you are running on backup power, you need to react fast to protect the systems or trigger other mitigation activities.

Here are some telemetry examples:

- Power supply state

- Backup power supply state

- Voltage

- Wattage

- Current

As infrastructure components have been used for many years, there are some agreed-upon standards for data structures and transmission. Let's look at those original standards now.

Common standards for infrastructure components

There are a few well-established standards that are used by infrastructure components that you may need to monitor. These include the following:

- **Syslog**: Syslog has been around since the 1980s and is very common in infrastructure components. Created as part of the **Sendmail project** by Eric Allman, it was quickly adopted and became the standard logging solution on Unix-like platforms. It is very popular because of its ease of use. To use Syslog, you need a client available to receive the data, and each device needs to be configured to send data there. Common clients include RSyslog and Syslog-ng, and the OpenTelemetry Collector also supports this protocol.

    The Syslog message format provides a structured framework that has allowed organizations to provide vendor-specific extensions. Contributing to its success and longevity, most modern observability tooling providers still supply an interface to receive Syslog messages. The logs can then be accessed and analyzed alongside other system and application telemetry.

- **Simple Network Management Protocol** (**SNMP**): Forming part of the original Internet Protocol suite defined by the **Internet Engineering Task Force** (**IETF**), SNMP is commonly used in networking infrastructure. A lot of the protocol is not of interest for observability, but **SNMP Traps** allow devices to inform the manager about significant events.

    SNMP provides a common mechanism for network devices to relay management and, specifically in the context of this chapter, monitoring information within single and multi-vendor LAN or WAN environments. It is different from other telemetry receivers as it requires more specific knowledge of the devices on the network, and specific configurations for the metrics to be collected. Here are some examples of data that can be collected from SNMP:

Data Type	Example Metrics Collected
Network data	Processes
	Uptime
	Throughput
Device data	Memory usage
	CPU usage
	Temperature

Table 2.8 – Example SNMP Trap information

You may encounter other formats out in the wide world of engineering. We have covered a lot of the common formats here and have hopefully given you an indication of the types of information you will need to help you work with telemetry in Grafana. Grafana will handle just about whatever you can throw at it. Knowing what's important and preparing for that will help you when you're building your visualizations and alerts on that data. Now, let's quickly recap what we've covered in this chapter.

Summary

In this chapter, we explored the foundations that modern observability is built on. This will serve as easy reference and support for future chapters in this book and your own projects. First, we looked at the common log formats and their examples, which will assist us in *Chapter 4, Looking at Logs with Grafana Loki*. Then, we took a closer look at metrics, their differing types, some example protocols, and best practices to consider when designing metric-based observability. What we covered here will help with *Chapter 5, Monitoring with Metrics Using Grafana Mimir and Prometheus*. We then moved on to traces and spans, where we looked at current protocols and some best practices to consider when building an efficient and effective trace-based observability platform. This section lays the groundwork for *Chapter 6, Tracing Technicalities with Grafana Tempo*. After looking at the telemetry of observability, we learned about application instrumentation, which we will see more of in *Chapter 3, Setting Up a Learning Environment with Demo Applications*, and later chapters where we go into specifics with logs, metrics, and traces. Lastly, we considered some of the more traditional infrastructure telemetry.

With the overview of application and infrastructure instrumentation complete, we can now start playing with logs, metrics, and traces. In the next chapter, we will get our learning environment up and running.

3

Setting Up a Learning Environment with Demo Applications

This chapter guides you through setting up a **learning environment** for the practical examples throughout this book and for independent experimentation. As Grafana Labs offers a free tier of their cloud service, we will use this for the storage and searching of data. To produce rich, useful data, we will use the **OpenTelemetry** demo application. This demo application deploys the services needed to run the OpenTelemetry Astronomy Shop locally. These applications are written in many different languages and are instrumented to produce metrics, logs, and distributed traces. This application will help you interact directly (and via load generators) with real applications and see observability telemetry in real time in the Grafana Labs instance.

In this chapter, we're going to cover the following main topics:

- Introducing Grafana Cloud
- Installing the prerequisite tools
- Installing the OpenTelemetry Demo application
- Exploring telemetry from the demo application
- Troubleshooting your OpenTelemetry demo application

Technical requirements

We assume you are working on at least Windows 10 version 2004, macOS version 11, or a relatively recent installation of Linux (e.g., Ubuntu 20.10 or later); earlier versions are not supported. We will be using the OpenTelemetry Collector version 0.73.1 and the OpenTelemetry Demo version 0.26.0. Full installation instructions for these components are provided in this chapter.

All commands and configuration files to complete these steps are included in the GitHub repository at `https://github.com/PacktPublishing/Observability-with-Grafana/tree/main/chapter3`. You'll find the *Code in Action* videos for this chapter at `https://packt.link/GNFyp`.

Introducing Grafana Cloud

Grafana Cloud is a hosted platform-as-a-service observability tool. It offers the ability to quickly create infrastructure to receive and store log, metric, and trace data as well as tools to visualize the data. We are using Grafana Cloud to reduce the technical skill needed to engage with this book, although all the open source components from Grafana that we will introduce can be deployed locally as well.

Grafana Cloud offers free access. In this chapter, we will first set up a cloud account and familiarize ourselves with the administration and use of the tools.

> **Important note**
> Grafana Labs introduces changes on a regular basis. The information and screenshots in this chapter are based on Grafana version 10.2, which was released in October 2023.

Setting up an account

Creating a free Grafana Cloud account is simple. To do this, follow these steps:

1. Go to `https://www.grafana.com` and click on **Create Free Account**.

2. Sign up using any of the following:

    A. A Google account

    B. A GitHub account

    C. A Microsoft account

    D. An Amazon account

    E. An email address

3. Choose a team URL. This is the URL used to access your Grafana instance. In our sample setup, we chose `observabilitywithgrafana`:

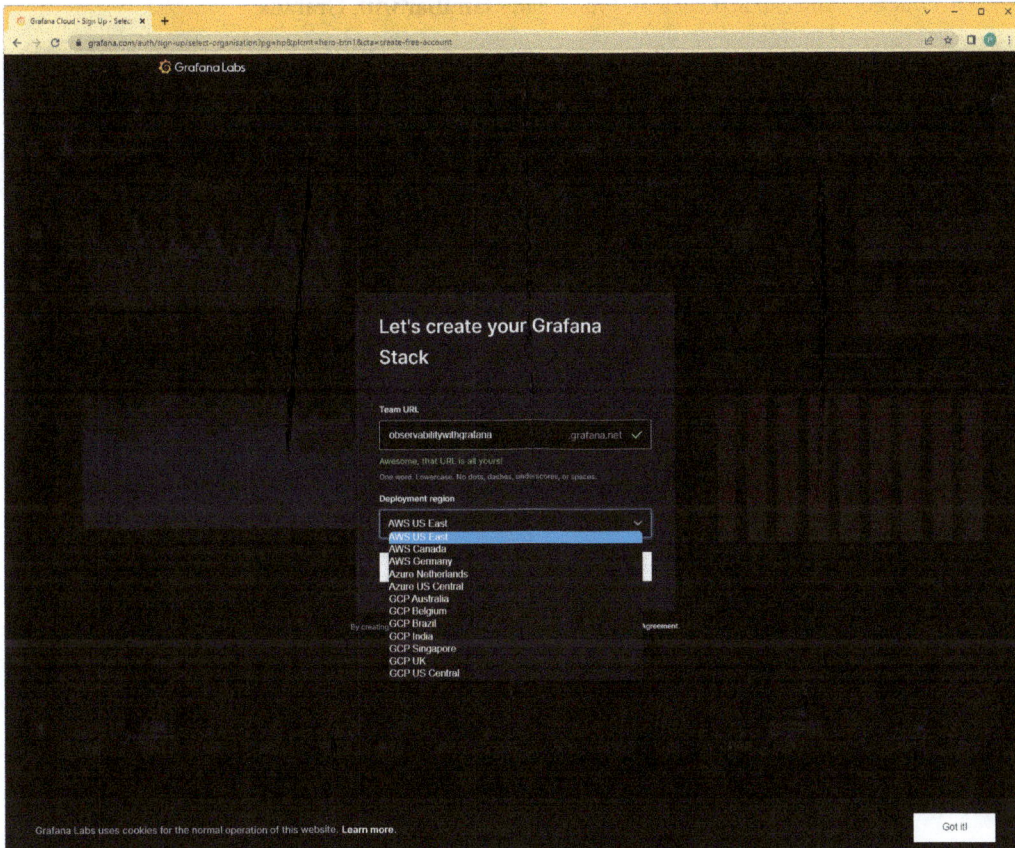

Figure 3.1 – Creating a Grafana stack

4. Choose a deployment region from the available list and then wait for a few minutes for your Grafana account to be created.

5.   When the account creation is completed, you will see a screen titled **GET STARTED**, as shown in the following screenshot. We encourage you to explore the screens in Grafana, but for this introduction, you can just click on **I'm already familiar with Grafana**:

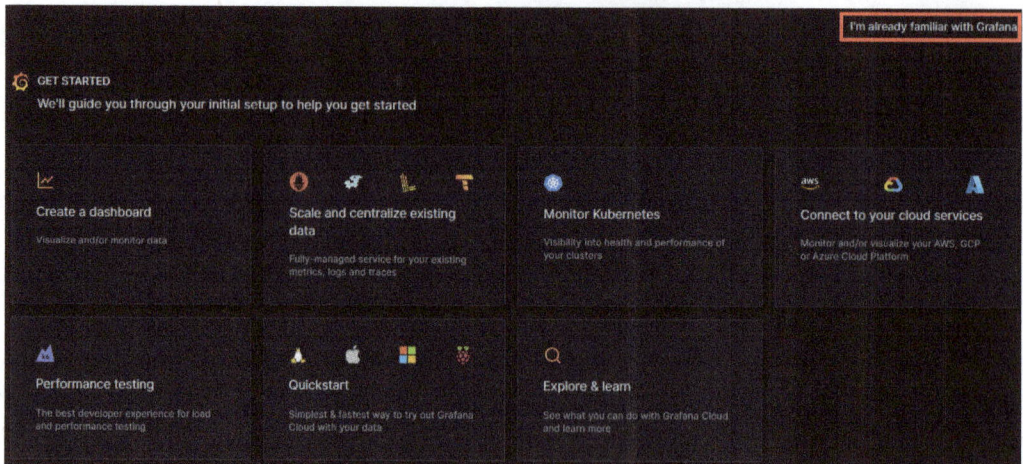

Figure 3.2 – Get started with Grafana

Then click on the Grafana icon in the top-left corner to return to the home screen, and finally, click on **Stacks**, which you will find just under the welcome message:

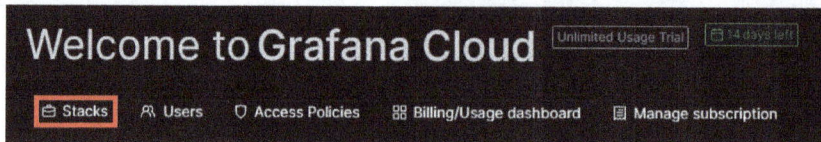

Figure 3.3 – Welcome to Grafana Cloud

6.   This will take you to the Grafana Cloud Portal, which will look like this.

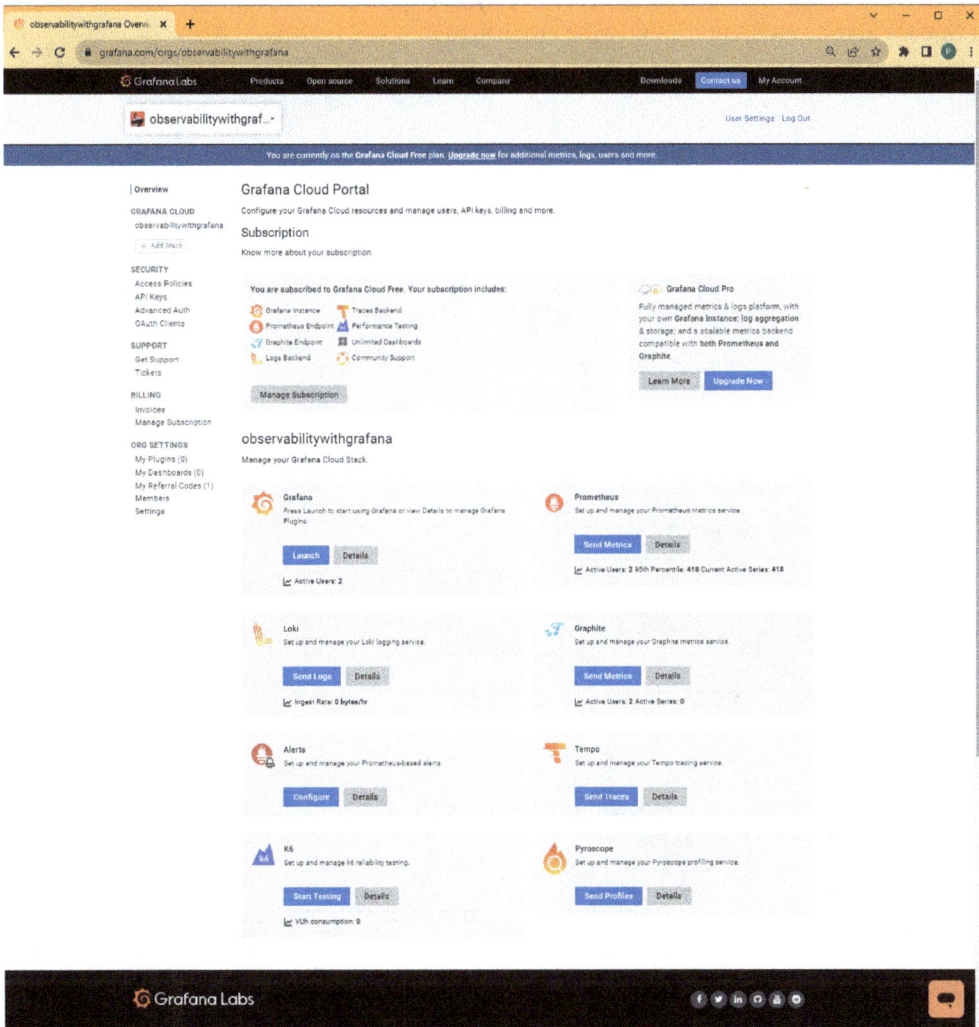

Figure 3.4 – The Grafana Cloud Portal

Before we set up the local data-producing environment, let's take a few moments to explore the basics of the Grafana Cloud Portal and your new Grafana Stack.

Exploring the Grafana Cloud Portal

The home page of the portal shows your subscription and billing information, as well as details about your single Grafana Stack. When you first sign up for a **Cloud Free** subscription, you will be granted access to a 14-day trial for Cloud Pro. To access your instance of Grafana, you need to click on the **Launch** button in the **Grafana** section. This gives you access to view data you send to Grafana Cloud.

A full Grafana **Stack** consists of the following installations:

- **Visualization**:

  - A Grafana instance

- **Metrics**:

  - A Prometheus remote write endpoint

  - A Graphite endpoint

  - A Mimir backend storage instance

- **Logs**:

  - A Loki endpoint

  - A Loki backend storage instance

- **Traces**:

  - A Tempo endpoint

  - A Tempo backend storage instance

- **Alerting**:

  - An Alerting instance for managing Prometheus-based alerts

- **Load testing**:

  - A k6 cloud runner

- **Profiling**:

  - A Pyroscope endpoint

  - A Pyroscope backend storage instance

Let's take a look at some of the sections you can access in your portal:

- **Security**: In this section, you can manage **access policies** and **access tokens**, **Open Authorization (OAuth)**, **Security Assertion Markup Language (SAML)**, and **Lightweight Directory Access Protocol (LDAP)**.

  Access policies are scoped to a **realm**; a realm can cover a specific stack, or it can cover the entire organization.

- **Support**: This section gives access to support from the community via **Community Forums**, and from Grafana Labs via **Support Ticket** and **Chat with Support**.

- **Billing**: The **Billing** section is where invoices and subscriptions are managed. A Cloud Free subscription provides a monthly limit of 10,000 metrics, 50 GB logs and 50 GB traces, 50 GB profiles, as well as three **Incident Response & Management (IRM)** users and 500 k6 **virtual user hours (VUh)**. This subscription is plenty for the demo applications we will be using during the book and similar small demo, personal-use installations. For larger installations, Grafana Cloud also offers two additional tiers, Cloud Pro and Cloud Advanced. These tiers give access to different key features, with Cloud Advanced being geared toward enterprise-level installations. Grafana then bills based on ingest and active users in different areas. Grafana provides a very easy-to-use monthly cost estimation tool:

Estimate your monthly costs

Use this tool to estimate your monthly costs based on anticipated usage.
This will not impact your monthly subscription as you only pay for what you use.

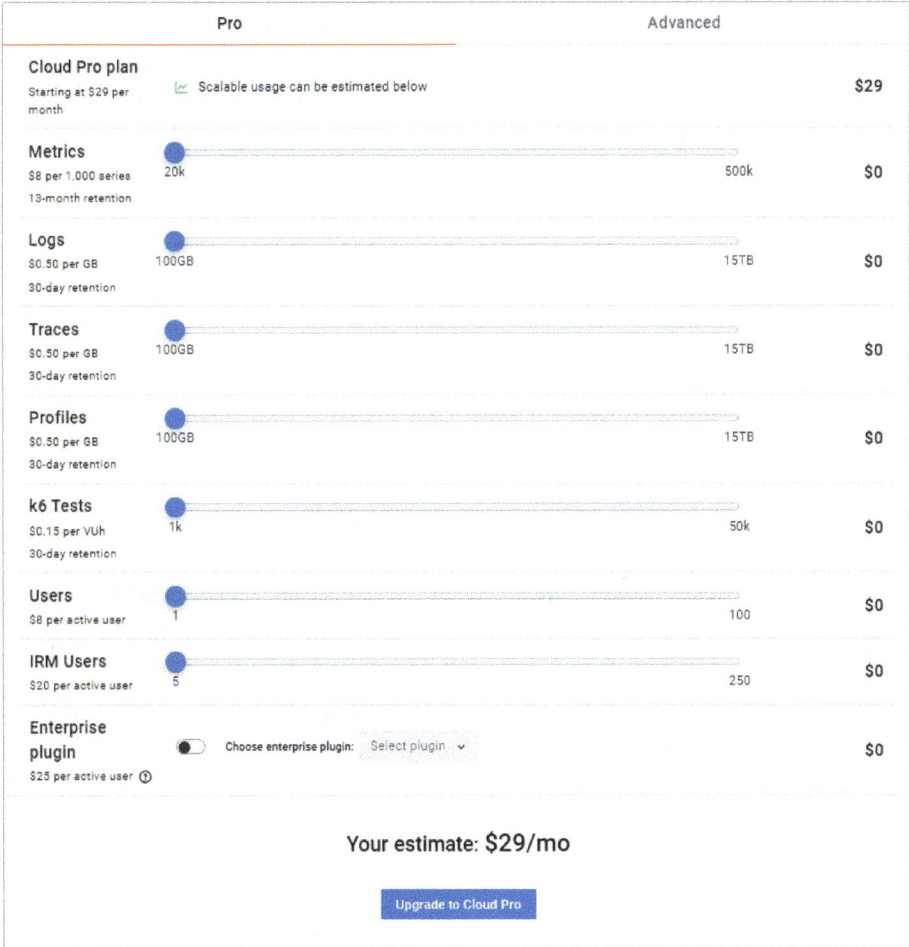

	Pro	Advanced
Cloud Pro plan Starting at $29 per month	📈 Scalable usage can be estimated below	$29
Metrics $8 per 1,000 series 13-month retention	20k ●——————————————————— 500k	$0
Logs $0.50 per GB 30-day retention	100GB ●——————————————————— 15TB	$0
Traces $0.50 per GB 30-day retention	100GB ●——————————————————— 15TB	$0
Profiles $0.50 per GB 30-day retention	100GB ●——————————————————— 15TB	$0
k6 Tests $0.15 per VUh 30-day retention	1k ●——————————————————— 50k	$0
Users $8 per active user	1 ●——————————————————— 100	$0
IRM Users $20 per active user	5 ●——————————————————— 250	$0
Enterprise plugin $25 per active user ⓘ	⬤ Choose enterprise plugin: Select plugin ⌄	$0

Your estimate: $29/mo

Upgrade to Cloud Pro

Figure 3.5 – Grafana cost estimator

With a Cloud Free subscription, you will only have access to a single stack. With a subscription to **Cloud Pro** or **Cloud Advanced**, you will be able to create multiple stacks in your account. These stacks can be in different regions.

- **Org Settings**: Here, you can manage the users who can access Grafana Cloud. You can also update your avatar, organization name, and any plugins or dashboards you have shared with the community.

Now that you have explored the Grafana Cloud Portal and have an overview of managing a Grafana Cloud account, let's explore the Grafana instance in your cloud stack.

Exploring the Grafana instance

The main way to interact with the tools provided by the Grafana Stack is to access the **Grafana instance**. This instance provides visualization tools to view the data that is collected by the rest of the Grafana stack or other linked tools. To access your Grafana instance, you can click on the **Launch** button in your Cloud Portal:

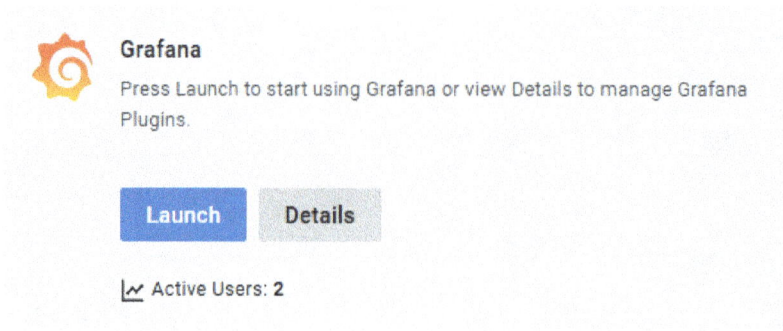

Figure 3.6 – Launching Grafana

To access it more directly, you can use the direct URL based on the team name selected during account creation. For instance, our sample account has the URL `https://observabilitywithgrafana.grafana.net`.

Once you have accessed Grafana, you will be greeted with a home page that shows your current usage. Navigation is done using the menu available at the top left of the page:

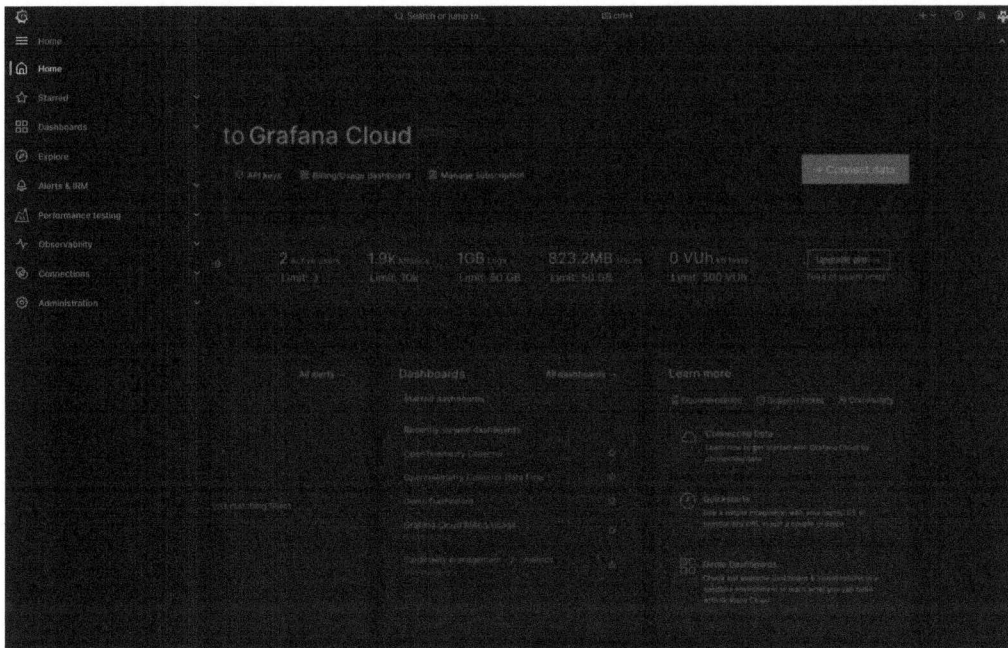

Figure 3.7 – The main navigation panel in Grafana

Let's talk through the different sections available in the menu shown in the preceding figure.

Dashboards and Starred dashboards

The **Dashboards** section allows you to create, import, organize, and tag dashboards in your Grafana installation. Under **Dashboards**, you can also access **Playlists**, **Snapshot**, **Library panels**, and **Reporting**:

- The **Playlists** subsection lets you manage groups of dashboards that are displayed in a sequence, for instance, on TVs in an office.

- The **Snapshot** tab gives you the ability to take a snapshot of a dashboard with the data to share. This will give the recipients the ability to explore the data much better than by sharing a simple screenshot.

- **Library panels** provides reusable panels that can be used in multiple dashboards to aid in offering a standardized UI.

- Finally, **Reporting** allows you to automatically generate PDFs from any dashboard and send them to interested parties on a schedule.

The **Starred** dashboards section lets you customize the UI by saving dashboards you frequently use at the top of the menu. We will explore using dashboards more fully in *Chapter 8*.

Explore

Explore is the way to get direct access to the data that is connected to the Grafana instance. This is the foundation of building custom dashboards and exploring data to understand the current state of the system. We will look at this briefly later in this chapter in the section titled *Exploring telemetry from the demo application*, and in a lot more detail in *Chapters 4, 5*, and *6*.

Alerts & IRM

The **Alerts & IRM** section contains Grafana's **Alerting, OnCall**, and **Incident** systems for IRM, as well as machine learning capabilities and the SLO service. These can help with identifying anomalies in the system (for more details on this, you can refer to *Chapter 9*):

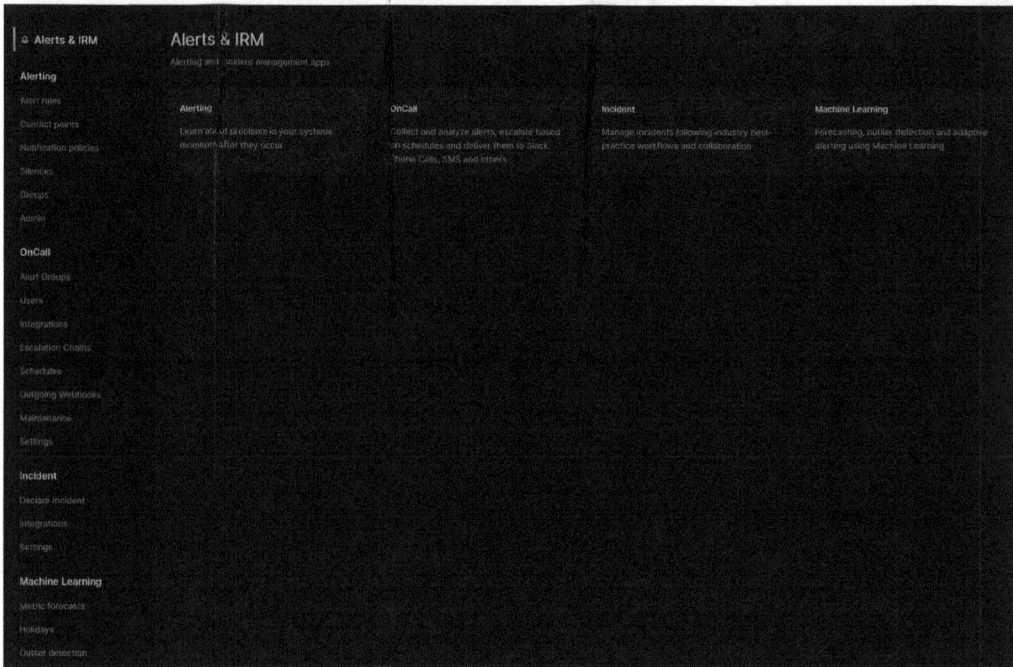

Figure 3.8 – Alerts & IRM

Alerting, or Alertmanager, allows teams to manage rules for when an alert should be triggered. It also gives teams control over the method of notification used for the alert and whether there are any schedules for when to alert. Grafana **Alerting** is geared toward notifying an individual team of alerts in their system, and it has been battle-tested by many organizations.

OnCall and **Incident** form Grafana's IRM toolkit. This gives organizations the ability to receive notifications from many monitoring systems and manage schedules and escalation chains, as well as giving high-level visibility of sent, acknowledged, resolved, and silenced alerts. The **Incident** section gives organizations the ability to start incidents. These follow predefined processes for involving the right people and communicating with stakeholders. Once the incident is initiated, Grafana can record key moments during the incident process ready for any post-incident processes. Both **OnCall** and **Incident** are aimed at organizations that need a central hub to manage the technical aspects of problem and incident management. *Chapter 9* will explore these tools in greater detail and show you how to configure them.

Performance testing

The **Performance testing** area is where the k6 performance testing tool is integrated with the Grafana UI, allowing teams to manage tests and projects. k6 allows your teams to use the same performance testing tools in the CI/CD pipeline as they use to test against production. k6 will be covered in *Chapter 13*.

Observability

The **Observability** section brings together Kubernetes infrastructure and application monitoring, with synthetic tests used to simulate critical user journeys in your application alongside frontend **real user monitoring (RUM)** and **profiles** from Pyroscope. By combining these data sources, you gain an end-to-end view of the current performance of your products.

Connections

The **Connections** section is an administrative area to set up and manage connections from Grafana to the 160+ data sources and infrastructure components that it can connect and show data from. Some examples of available connections are Elastisearch, Datadog, Splunk, AWS, Azure, cert-manager, GitHub, Ethereum, and Snowflake. The page to configure these connections is shown in the following figure:

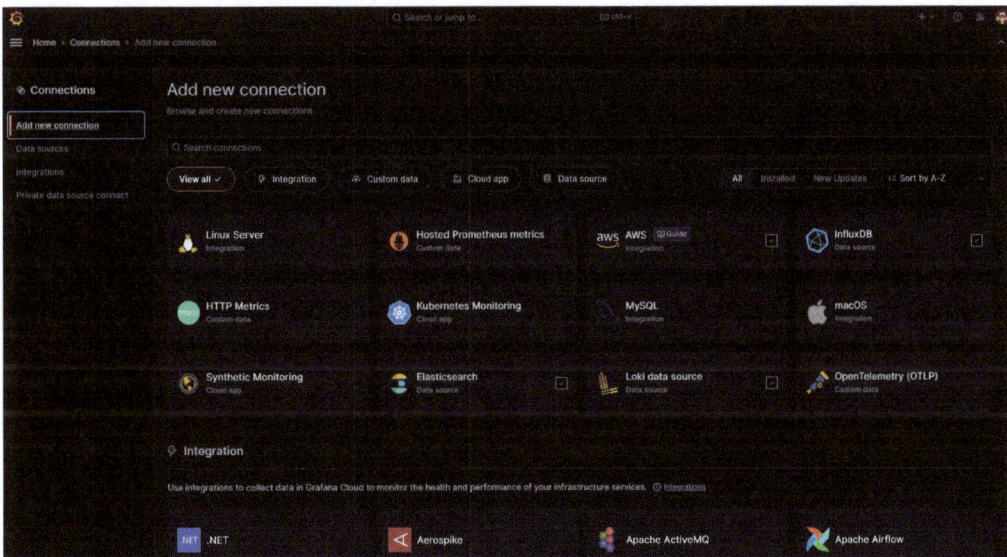

Figure 3.9 – Connections screen

Administration

The **Administration** panel allows for the management of many aspects of Grafana, from plugins and recorded queries to users, authentication, teams, and service accounts:

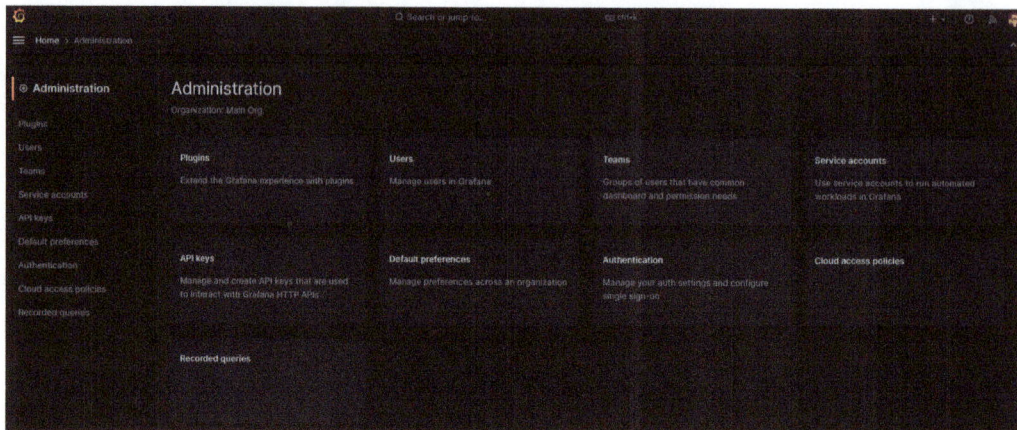

Figure 3.10 – Administration panel

Now that we've explored the Grafana Cloud Portal, let's prepare your local environment to send data.

Installing the prerequisite tools

Grafana is much more exciting with data! To build a realistic view of how Grafana works and can help your organization, we have chosen to install the OpenTelemetry demo application. This is a demonstration web store that sells telescopes and other observational equipment. We'll take you through the installation process to get this running locally on your machine.

But first, there are a few prerequisites that your local environment has, and they depend on the operating system you use.

In this section, we will explain how to install the following:

- Tools based on your operating system:

  - **Windows Subsystem for Linux version 2 (WSL2)**

  - macOS Homebrew

- Docker or Podman

- A single-node Kubernetes cluster

- Helm

Installing WSL2

WSL is a way of running a Linux filesystem and tools directly on Windows. This is used to give a standard set of commands and tools across operating systems. It is possible to run these systems outside of WSL, but the processes are much easier using WSL. Follow these steps to set up WSL2:

1. Open a PowerShell terminal or Windows Command Prompt as an administrator.

2. Run the following command to install WSL2:

   ```
   C:\Users\OwG> wsl --install
   ```

   This will install the Ubuntu distribution of Linux.

   You should see a message that Ubuntu is installing and then you will be prompted for a new UNIX username. If you see any other message, you may have a version of WSL installed already; please refer to the Microsoft website to address this.

3. Create a new user for your Ubuntu installation by entering a username and password. These do not need to match your Windows username and password.

4. Upgrade packages by running the following command:

   ```
   sudo apt update && sudo apt upgrade
   ```

   It is recommended to install the Windows Terminal to use WSL. This can be downloaded from the Microsoft Store: https://learn.microsoft.com/en-us/windows/terminal/install.

More detailed instructions on how to complete this installation process and troubleshoot any issues that may arise can be found at https://learn.microsoft.com/en-us/windows/wsl/install.

Installing Homebrew

Homebrew is a package management tool for macOS. To install it on your system, follow these steps:

1. Open a terminal window.

2. Run the following command to install Homebrew:

   ```
   $ /bin/bash -c "$(curl -fsSL https://raw.githubusercontent.com/
   Homebrew/install/HEAD/install.sh)"
   ```

   The script that runs will explain what it is doing and pause before continuing.

3. Run the following command to install wget (used later in the setup process):

   ```
   $ brew install wget
   ```

Detailed instructions on this installation can be found at https://brew.sh/.

Installing container orchestration tools

Docker and **Podman** are both **container orchestration tools**. Docker has been the de facto standard for about a decade. Podman is a newer tool. It was first released in 2018. The major functions of Docker are also offered in Podman, but support and information may be more difficult to find.

In 2021, Docker made licensing changes that became fully effective in January 2023. These licensing changes may impact readers using devices belonging to businesses. We will only provide instructions for Docker, but we will highlight the potential issues, so you are aware of them. The aim of providing a container orchestration tool is to run a single-node Kubernetes cluster on that tool.

Installing Docker Desktop

The following sections briefly describe how **Docker Desktop** can be installed on Windows, macOS, and Linux. Let's look at the Windows installation process first.

Windows installation

To install Docker Desktop on Windows, do the following:

1. Go to `https://docs.docker.com/desktop/install/windows-install/` and download the latest Docker installer `.exe` file.
2. Run the installation package. You will be prompted to enable WSL2. Select **Yes**.
3. After installation, start Docker Desktop.
4. Navigate to **Settings** | **General**. Select the **Use WSL2 based engine** checkbox.
5. Select **Apply & Restart**.
6. In your WSL terminal, validate the Docker installation by running the following command:

```
$ docker --version
Docker version 20.10.24, build 297e128
```

macOS installation

For the macOS installation of Docker Desktop, do the following:

1. Go to `https://docs.docker.com/desktop/install/mac-install/` and download the latest Docker installer for your Mac.
2. Run the installation package.
3. In your terminal, validate the Docker installation by running the following:

```
$ docker --version
Docker version 20.10.24, build 297e128
```

Linux installation

Linux users may consider using Podman Desktop as it has better native support for Linux environments. Installation instructions are available at `https://podman-desktop.io/docs/Installation/linux-install`.

However, if you do wish to use Docker, please follow the installation guide provided as the process will vary by distribution: `https://docs.docker.com/desktop/install/linux-install/`.

The full documentation can be found at `https://docs.docker.com/desktop/`.

With a base system that now supports containerization, let's set up a single-node Kubernetes cluster. This will be used to easily and repeatably install the OpenTelemetry demo application.

Installing a single-node Kubernetes cluster

There are a few different tools to run local Kubernetes clusters, including **Kubernetes in Docker (KinD)**, **Minikube**, and **MicroK8s**. However, we have chosen to use **k3d** due to the ease of installation across operating systems. Follow these steps:

1. Install k3d using the following command:

   ```
   $ wget -q -O - https://raw.githubusercontent.com/k3d-io/k3d/
   main/install.sh | bash
   ```

   Full details can be found at `https://k3d.io/stable/`.

2. Create a k3d cluster:

   ```
   $ k3d cluster create owg-otel-demo
   ```

3. Validate the cluster status:

   ```
   $ kubectl get nodes
   NAME                    STATUS    ROLES            AGE    VERSION
   k3d-owg-otel-demo-server-0    Ready    control-
   plane,master    13d    v1.25.7+k3s1
   ```

With a Kubernetes cluster installed, we now need a way to install applications easily. OpenTelemetry provides Helm charts, which allow us to deploy applications to Kubernetes; we will install Helm to use these.

Installing Helm

Helm is a package manager for Kubernetes clusters. A lot of infrastructure-level components are provided as Helm charts for installation, and OpenTelemetry is no exception. To install Helm, follow these steps:

1. Install Helm using the following command (details can be found at `https://helm.sh/docs/intro/install/`):

   ```
   $ wget -q -O - https://raw.githubusercontent.com/helm/helm/main/
   scripts/get-helm-3 | bash
   ```

2. Validate the Helm installation:

   ```
   $ helm version
   version.BuildInfo{Version:"v3.11.3",
   GitCommit:"323249351482b3bbfc9f5004f65d400aa70f9ae7",
   GitTreeState:"clean", GoVersion:"go1.20.3"}
   ```

With our local system ready to start running applications, we can now install the demo application provided by OpenTelemetry.

Installing the OpenTelemetry Demo application

As we need to send data from our local machine to our Grafana Cloud instance, we need to provide some credentials and tell our local machine where to send data. Before we install the demo application, we need to set up access tokens so we can send data to our Grafana Cloud stack.

Setting up access credentials

Access tokens allow the Open Telemetry application to send data securely to your Grafana Cloud stack. To set up an access token, we need to do the following:

1. In the Grafana Cloud Portal, select **Access Policies** in the **Security** section.

2. Click **Create access policy**.

3. Choose a name and display name, and set **Realm** to **All Stacks**.

4. Set the **Write** scope for metrics, logs, and traces and click **Create**.

5. On the new access policy, click **Add token**.

6. Give the token a name and set the expiration date. Click **Create**.

7. Finally, copy the token and save it securely.

Now that we've created our token, let's download the GitHub repository for this book and set it up.

Downloading the repository and adding credentials and endpoints

The Git repository contains configuration files for each of the chapters that will use live data. Downloading the repository is simple. Go to the Git repository: `https://github.com/PacktPublishing/`

`Observability-with-Grafana`. Click on the green **Code** button and follow the instructions to clone the repository. Once downloaded, the repository should look like this:

```
$ tree Observability-with-Grafana
Observability-with-Grafana
├── chapter3
│   └── OTEL-Collector.yaml
├── chapter4
│   └── OTEL-Collector.yaml
├── chapter5
│   └── OTEL-Collector.yaml
├── chapter6
│   └── OTEL-Collector.yaml
├── OTEL-Creds.yaml
├── OTEL-Demo.yaml
└── README.md
```

Figure 3.11 – Observability with Grafana repo

To set up the demo application, we need to add the token we saved in the previous section and the correct endpoints in the `OTEL-Creds.yaml` file. The information you will need is the following:

- **Password**: This is the token you saved earlier and is shared by each telemetry type
- **Username**: This is specific to each telemetry type
- **Endpoint**: This is also specific to each telemetry type

Let's take Loki as an example of how to get this information. In the Grafana Cloud Portal, do the following to get the correct information:

1. Click on the **Send Logs** button in the Loki box. At the top of the page is an information box that looks like this:

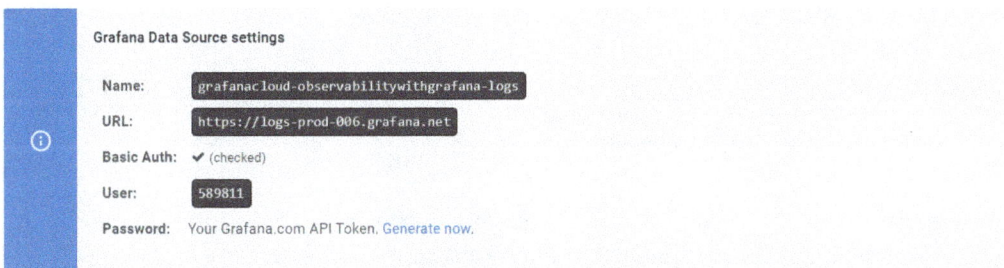

Grafana Data Source settings

Name:	`grafanacloud-observabilitywithgrafana-logs`
URL:	`https://logs-prod-006.grafana.net`
Basic Auth:	✔ (checked)
User:	`589811`
Password:	Your Grafana.com API Token. Generate now.

Figure 3.12 – Grafana Data Source settings

2.   The **User** field is used as the username.

3.   The **URL** field is used as the endpoint. This needs modifying:

   •   For logs, add `/loki/api/v1/push` to the end of the URL

   •   For metrics, add `/api/prom/push` to the end of the URL

   •   For traces, remove `https://` and add `:443` to the end of the URL

4.   Add this information into the relevant fields in the `OTEL-Creds.yaml` file.

With Loki completed, we can follow the same process for Prometheus (click **Send Metrics**) and Tempo (click **Send Traces**).

These instructions are repeated and expanded on in the `README.md` file in the repository. With the credentials saved, we are now ready to deploy OpenTelemetry.

Installing the OpenTelemetry Collector

With the credentials file updated, we can now install the Helm chart for the OpenTelemetry Collector. This is the component that will collect and transmit the data produced by the demo application. To proceed with the installation, perform the following steps:

1.   Add the OpenTelemetry repository:

```
$ helm repo add open-telemetry https://open-telemetry.github.io/
opentelemetry-helm-charts
```

2.   Install the collector with Helm:

```
$ helm install --version '0.73.1' --values chapter3/OTEL-
Collector.yaml --values OTEL-Creds.yaml owg open-telemetry/
opentelemetry-collector
NAME: owg-otel-collector
LAST DEPLOYED: Sun May 14 13:53:16 2023
NAMESPACE: default
STATUS: deployed
REVISION: 1
...
```

3.   Validate that the installation was successful:

```
$ kubectl get pods
NAME                                      READY   STATUS    RESTARTS   AGE
owg-otel-collector-opentelemetry-collector-6b8fdddc9d-
4tsj5    1/1      Running   0           2s
```

We are now ready to install the OpenTelemetry demo application.

Installing the OpenTelemetry demo application

The OpenTelemetry demo application is an example web store that sells telescopes and other tools to observe the universe. To install this demo application, follow these steps:

1. Install the demo application with Helm:

   ```
   $ helm install --version '0.26.0' --values owg-demo open-
   telemetry/opentelemetry-demo
   NAME: owg-otel-demo
   LAST DEPLOYED: Mon May 15 21:58:37 2023
   NAMESPACE: default
   STATUS: deployed
   REVISION: 1
   …
   ```

2. Validate that the installation was successful. This process may take a few minutes on the first install:

   ```
   $ kubectl get pods
   ```

   You should see about 19 Pods running.

3. Open a port for `frontendproxy` access:

   ```
   $ kubectl port-forward svc/owg-demo-frontendproxy 8080:8080 &
   ```

4. Open a port for browser spans:

   ```
   $ kubectl port-forward svc/owg-opentelemetry-collector 4318:4318
   &
   ```

5. Check you can access the OpenTelemetry demo application on `http://localhost:8080`:

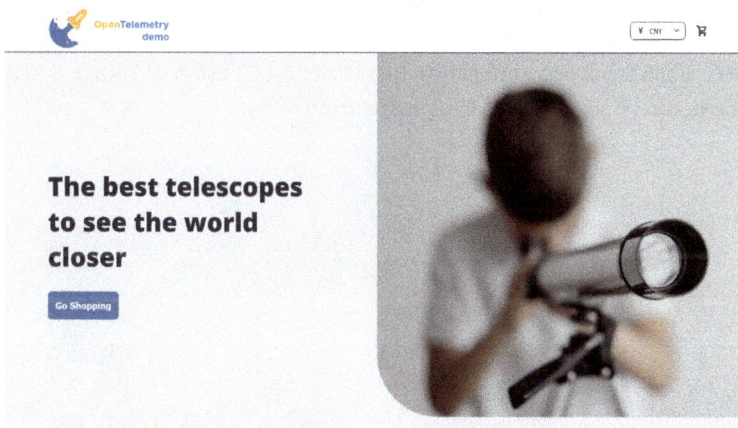

Figure 3.13 – OpenTelemetry demo application

6. Explore the OpenTelemetry demo application to generate some data. Take some time to add items to your cart and purchase them. While a load generator is part of the installation, it is worth getting familiar with the application as it will help put the telemetry generated in context.

Now that you have an application that produces data, let's start exploring that data.

Exploring telemetry from the demo application

Assuming the installation was successful, you will have data flowing into your Grafana instance. To view these, head to your Grafana instance, either from your team URL or by clicking **Launch** in the Cloud Portal.

On the front page, you should see information detailing your current or billable usage of logs, metrics, and traces. If everything is correct, these should all be greater than 0. It may take a few minutes for this to show up completely, sometimes up to an hour:

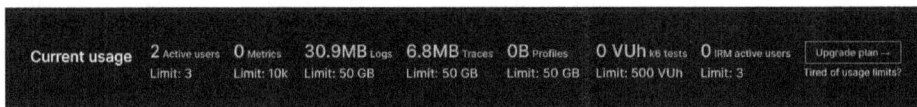

Figure 3.14 – Grafana Cloud usage

> **Important note**
>
> If you don't see usage for metrics, logs, and traces, you might need to troubleshoot. We have included some tips at the end of this chapter.

Click on the menu and select **Explore**. This will take you to the page to run queries on the telemetry from the demo application. The highest-level concept in Grafana is a **data source**. These are selectable just below the main menu. Each data source is a connection from the Grafana instance to either a component in the Grafana Stack or a connection to another tool. Each data source is self-contained. The data source menu can be seen in the following figure:

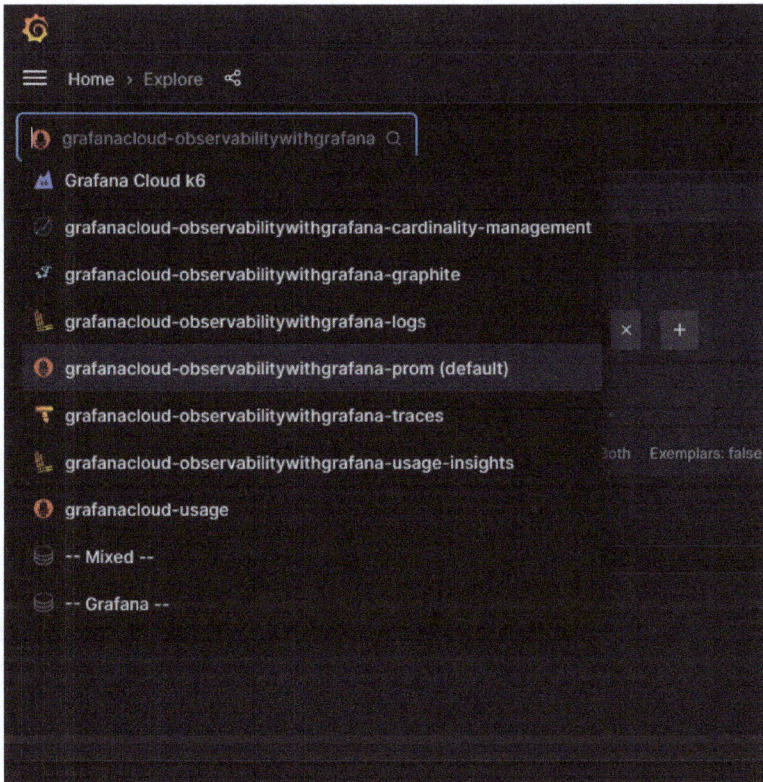

Figure 3.15 – Data source menu

You will see that there are data sources for **logs**, **prom** (Prometheus metrics), and **traces**. There are other data sources, but for this chapter, we will only concentrate on these three, as they are the data sources related to the Grafana Stack we have provisioned.

These data sources relate to our Loki, Mimir, and Tempo components, used to collect logs, metrics, and traces, respectively. We'll discuss these next.

Logs in Loki

Loki stores the log data produced by the demo application. Let's have a quick look at the interface for a Loki data source and see some of the log data the demo application is producing.

Select the `grafanacloud-<team name>-logs` data source. In **Label filters**, add **exporter = OTLP** and click **Run query**. You should now see a screen like this:

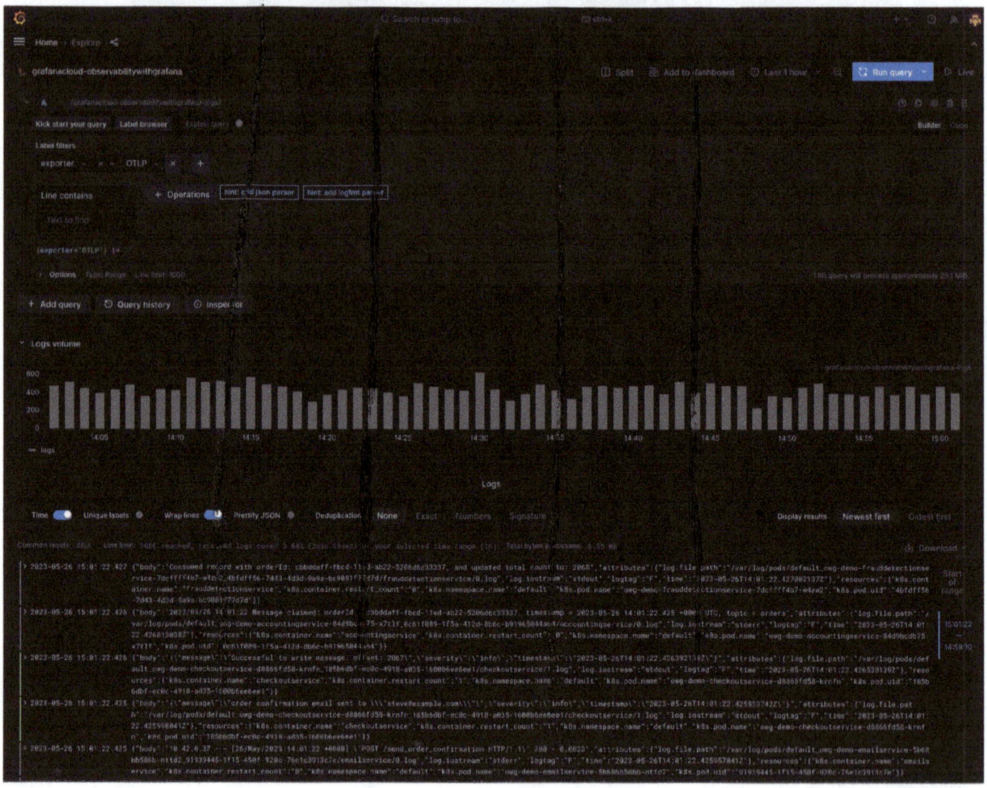

Fig 3.16 – Reviewing logs

The upper area of the screen is where you enter and modify your query. In the middle of the screen, you will see the volume of logs returned by your search over time. This can provide a lot of contextual information at a glance. The lower part of the screen displays the results of your query:

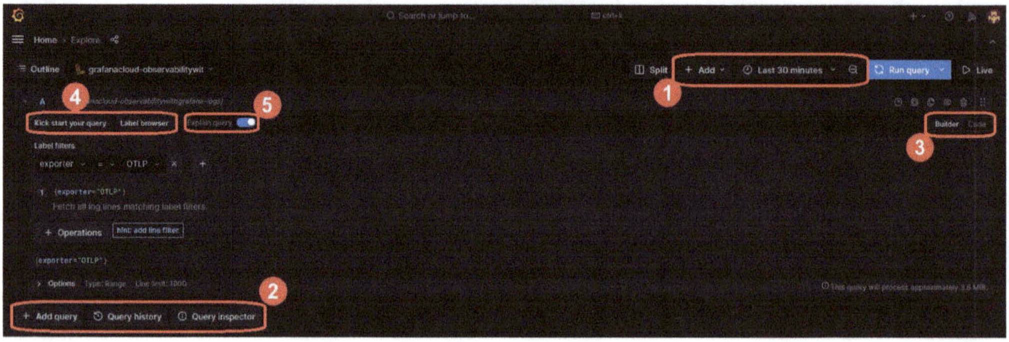

Figure 3.17 Query panel details

Let's break the preceding screenshot down into its components:

1. You can select the time range and add the query to various tools by clicking on **Add**.

2. Using the buttons at the bottom of this section, you can use **Add query** to produce a more complex analysis of the data. **Query history** lets you review your query history. **Query inspector** helps to understand how your query is performing.

3. The **Builder** or **Code** selector allows you to switch between the UI-focused query builder you see by default and a query language entry box, which is useful for directly entering queries.

4. The **Kick start your query** button will give you some starter queries and the **Label browser** button will let you explore the labels currently available in your collected log data.

5. Finally, the **Explain query** slider will give you a detailed explanation of each step of the query you currently have.

Let's take a look at the components in the query result panel:

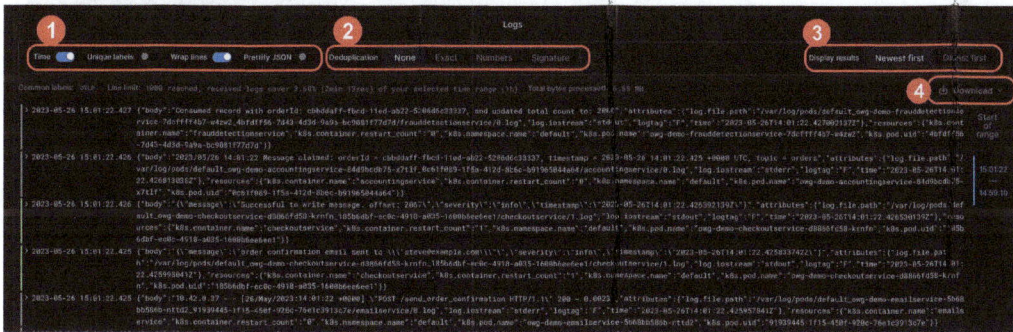

Figure 3.18 – Query result panel details

These components can be used as follows:

1. You can customize the display of this data to show the timestamp, highlight any unique labels, wrap lines, or prettify JSON.

2. You can also apply deduplication.

3. You can choose to view the oldest or newest log events first.

4. Finally, you can download the data as .txt or .json.

We will look at these features in more depth in *Chapter 4* and introduce the LogQL query language.

Now that you have familiarized yourself with the query panels for logs in a Loki data source, let's see how querying metrics from a Mimir data source is very similar.

Metrics in Prometheus/Mimir

Grafana uses similar repeating blocks for querying data. In this section, we will explore how to query metrics.

Select the **grafanacloud-<team name>-prom** data source. In the **Metric** dropdown, select **kafka_consumer_commit_rate** and leave **Label filters** blank. Click on **Run query** and you should see a screen like this:

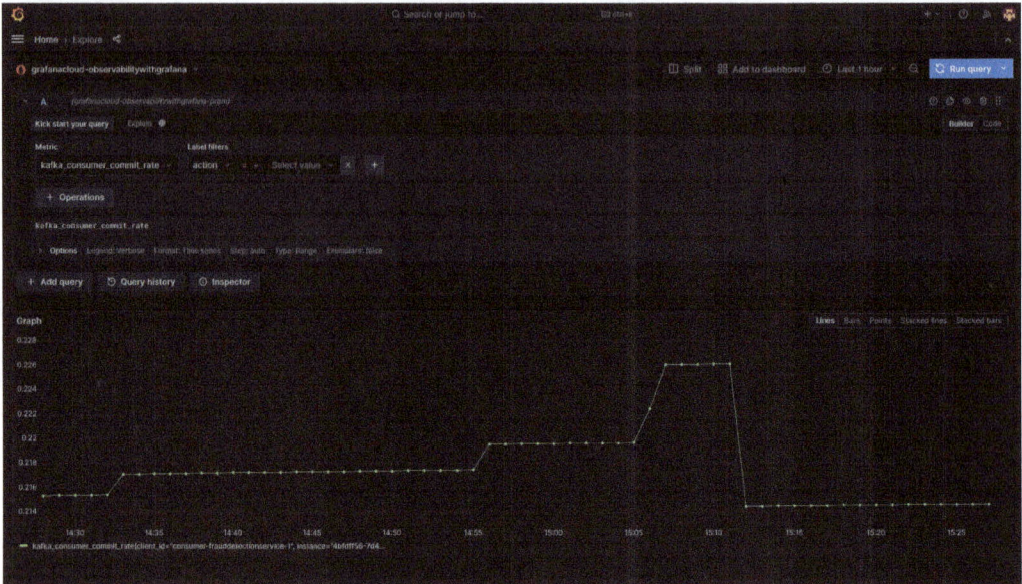

Figure 3.19 – Kafka consumer commit rate

Similar to Loki, we have a query section at the top, although the structure is slightly different. The major difference is the **Options** section, which controls how the data is presented. **Legend** manages the graph's legend display. The **Format** option allows you to select between a **Time series** representation (the graph shown in the preceding screenshot); a **Table** representation, showing each value of each time series; and a **Heatmap** representation. The **Type** selector allows you to select either a range of time, the latest instant of each time series, or both. The **Exemplars** slider will show trace data that is linked to each metric.

The bottom section shows the data, with options to display it in different ways.

To see how Grafana represents multiple plots, from the **Metric** dropdown, select **process_runtime_ jvm_cpu_utilization_ratio** and click **Run query**. You should see a chart like this:

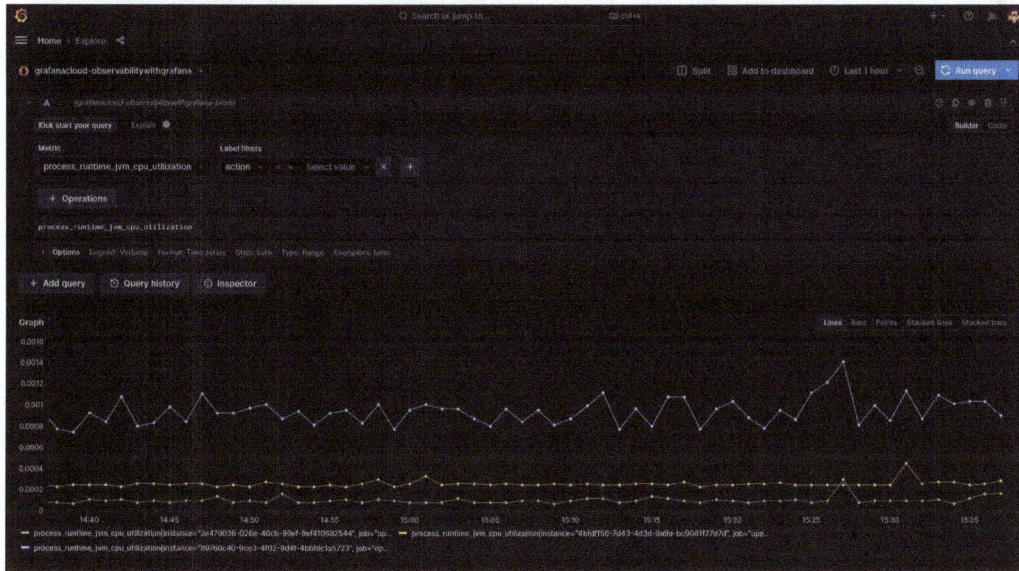

Figure 3.20 – JVM CPU utilization

Each of the plotted lines in the chart in the preceding screenshot represents the CPU utilization of a different service running in the demo application. We will look at these features in more depth in *Chapter 5* and introduce the PromQL query language.

A Tempo trace data source is very similar to a Loki log data source and a Mimir metric data source. Let's look at this now.

Traces in Tempo

Traces use a little bit more querying, as we have to find and select an individual trace, and then show the spans in the trace in detail.

Select the **grafanacloud-<team name>-traces** data source. For **Query type**, choose **Search**, for **Service Name**, select **checkoutservice**, and click **Run query**. You should see a screen like this:

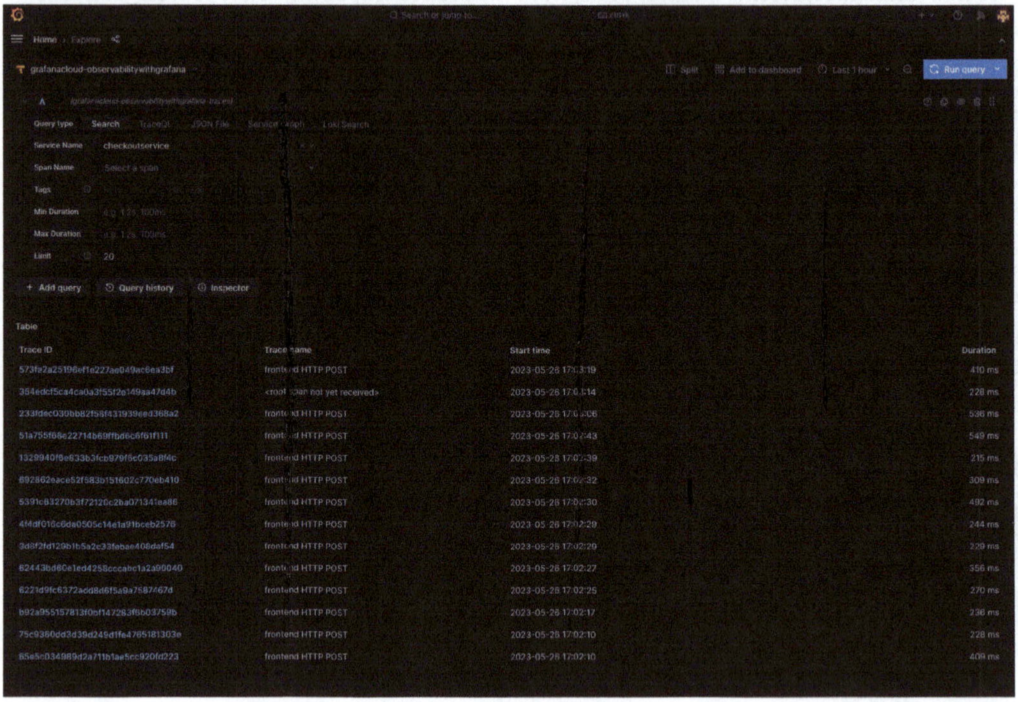

Figure 3.21 – Selecting a trace

Click on one of the **Trace ID** links and you will see a screen like this:

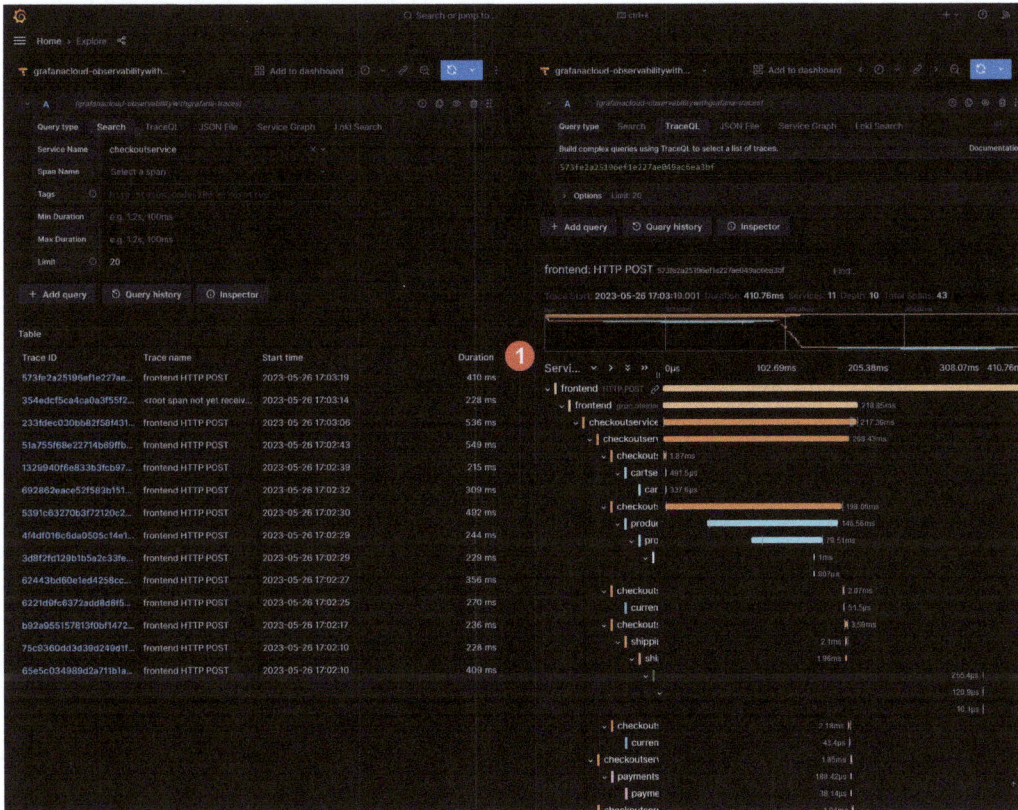

Figure 3.22 – Trace view

This view shows the split view that Grafana offers. This can be accessed from any data source and can provide the ability to use linked data points to move between logs, metrics, and traces to see the same point in time or event in each of the telemetry types. If you select the bar between the panels, indicated by the number **1** in the preceding screenshot, you can make the trace view bigger by clicking and dragging.

Looking through the trace, each line represents a span with information on which service was involved alongside specific information from that service. For example, **checkoutservice** will show you the count of items in the cart and the shipping amount. The bars with each span show the total time that span took and are stacked to represent the relative start and end times of the span. You might see that **productcatalogueservice** takes up a significant portion of the time in the transaction in *Figure 3.21*.

Adding your own applications

It is possible to add your own application to this demo installation. The OpenTelemetry Collector has been deployed with receivers available for OTLP-formatted data on port `4317` for gRPC data and `4318` for HTTP(s) data. To deploy your application, it will need to be packaged as a container and then just follow a standard deployment mechanism for a Kubernetes deployment. Please do note that k3d is opinionated and by default uses Flannel and Traefik for networking and ingress, respectively.

Now that we've understood this demo application, let's look at some troubleshooting tips that will help you if you're facing any issues while setting it up or using it.

Troubleshooting your OpenTelemetry Demo installation

There are various things that could go wrong while using this application, so this section is not necessarily exhaustive. We will assume that all the prerequisite tools were installed correctly.

Let's first take a look at the correct formatting of Grafana credentials.

Checking Grafana credentials

If the OpenTelemetry Collector is giving authentication errors, or you are not receiving data from it, it is likely that the credentials are not formatted correctly. These details are entered into the `OTEL-Creds.yaml` file from this book's Git repository.

The Grafana credentials should be formed like this:

- **Username**: Number, typically six digits.

- **Password**: API token, which is a long string of letters and numbers. When troubleshooting the token there are some important considerations:

  - The same token can be shared by all the exporters.

  - The **Access policy** associated with the API token needs to be able to write logs, metrics and traces. This can be checked in the Grafana Cloud portal.

- **Endpoint**: The URL that the OpenTelemetry Collector will push data to. These are the endpoints that are provided by the Grafana Cloud installation you set up in an earlier section, *Introducing Grafana Cloud*.

  The following are sample endpoints. Yours will be different, but they will be similarly formed:

  - `loki`: `https://logs-prod-006.grafana.net/loki/api/v1/push`

  - `prometheusremotewrite`: `https://prometheus-prod-13-prod-us-east-0.grafana.net/api/prom/push`

  - `otlp` (tempo): `tempo-prod-04-prod-us-east-0.grafana.net:443`

> **Important note**
> The OTLP endpoint for Tempo is different from the others.

If you notice that you have made a mistake, you will need to use `helm upgrade` to deploy the changes you make to the `OTEL-Creds.yaml` file:

```
$ helm upgrade --version '0.73.1' --values chapter3/OTEL-Collector.
yaml --values OTEL-Creds.yaml owg open-telemetry/opentelemetry-
collector
```

You can also use `helm uninstall` and then use the original instructions to reinstall.

The most common problem is with the access credentials, but sometimes you will need to look at the collector logs to understand what is happening. Let's see how to do this now.

Reading logs from the OpenTelemetry Collector

The next place to investigate is the logs from the OpenTelemetry Collector.

Kubernetes allows you to directly read the logs from a Pod. To do this, you need the full Pod name, which you can get using the following command:

```
$ kubectl get pods --selector=app.kubernetes.io/instance=owg
NAME                                    READY   STATUS     RESTARTS    AGE
owg-opentelemetry-collector-567579558c-
std2x    1/1     Running    0              6h26m
```

In this case, the full Pod name is `owg-opentelemetry-collector-567579558c-std2x`.

To read the logs, run the following command:

```
$ kubectl logs owg-opentelemetry-collector-567579558c-std2x
```

Look for any warning or error-level events, which should give an indication of the issue that is occurring.

Sometimes, the default logging does not give enough information. Let's see how we can increase the logging level.

Debugging logs from the OpenTelemetry Collector

If the standard logs are not enough to resolve the problem, the OpenTelemetry Collector allows you to switch on **debug logging**. This is typically verbose, but it is very helpful in understanding the problem.

At the end of `OTEL-Creds.yaml`, we have included a section to manage the debug exporter. The verbosity can be set to `detailed`, `normal`, or `basic`.

For most use cases, switching to `normal` will offer enough information, but if this does not help you to address the problem, switch to `detailed`. Once this option is changed, you will need to redeploy the Helm chart:

```
$ helm upgrade owg open-telemetry/opentelemetry-collector -f OTEL-
Collector.yaml
```

This will restart the Pod, so you will need to get the new Pod ID. With that new ID, you can look at the logs; as we saw in the previous section, you should have a lot more detail.

Summary

In this chapter, we have seen how to set up a Grafana Cloud account and the main screens available in the portal. We set up our local machine to run the OpenTelemetry demo application and installed the components for this. Finally, we looked a the data produced by the demo application in our Grafana Cloud account. This chapter has helped you set up an application that will produce data that is visible in your Grafana Cloud instance, so you can explore the more detailed concepts introduced later with real examples.

In the next chapter, we will begin to look in depth at logs and Loki, explaining how best to categorize data to give you great insights.

Part 2: Implement Telemetry in Grafana

This part of the book will take you through the different telemetry sources, explaining what they are, when to use them, and problems to watch out for. You will look at integrations with major cloud vendors: AWS, Azure, and Google. This part will also investigate real user monitoring with Faro, profiling with Pyroscope, and performance with k6.

This part has the following chapters:

- *Chapter 4, Looking at Logs with Grafana Loki*

- *Chapter 5, Monitoring with Metrics Using Grafana Mimir and Prometheus*

- *Chapter 6, Tracing Technicalities with Grafana Tempo*

- *Chapter 7, Interrogating Infrastructure with Kubernetes, AWS, GCP, and Azure*

4

Looking at Logs with Grafana Loki

In this chapter, we will get hands-on experience with **Grafana Loki**. We will learn how to use **LogQL**, which is the language used for querying Loki, how to select and filter log streams, and how to use the operators and aggregations available. This will give you the tools to extract the data in appropriate ways for your dashboard visualizations and alerts. We will review the benefits and drawbacks of the log format and how it impacts your use of Loki. To fully explore the benefits of Loki, we will explore the architecture and where it can be scaled for performance. To finish, we will look at advanced areas of LogQL, such as labels and transformations, and other tips and tricks to expand your use of Loki.

We will cover the following main topics in this chapter:

- Introducing Loki
- Understanding LogQL
- Exploring Loki's architecture
- Tips, tricks, and best practices

Technical requirements

In this chapter, you will work with LogQL using the Grafana Cloud instance and demo you set up in *Chapter 3*. The full LogQL language documentation can be found on the Grafana website at `https://grafana.com/docs/loki/latest/logql/`. Loki is in active development, so it's worth checking for new features frequently.

You'll find the code for this chapter in the GitHub repository at `https://github.com/PacktPublishing/Observability-with-Grafana/tree/main/chapter4`. You'll find the *Code in Action* videos for this chapter at `https://packt.link/aB4mP`.

Updating the OpenTelemetry demo application

First, let's improve the logging for our demo application. For this chapter, we have provided an updated `OTEL-Collector.yaml` file with additional Loki log labels in the `chapter4` folder in the GitHub repository. These instructions assume you have already completed the demo project setup in *Chapter 3*. Full details on this process are available in the GitHub repository in the *Chapter 4* section of the `README.md` file.

To upgrade the OpenTelemetry Collector, follow these steps:

1. Upgrade the collector with Helm:

    ```
    $ helm upgrade --version '0.73.1' --values chapter4/OTEL-
    Collector.yaml --values OTEL-Creds.yaml owg open-telemetry/
    opentelemetry-collector
    NAME: owg-otel-collector
    LAST DEPLOYED: Sun Apr 25 12:15:03 2023
    NAMESPACE: default
    STATUS: deployed
    REVISION: 2
    ...
    ```

2. You can validate that the upgrade was successful with this command:

    ```
    $ kubectl get pods --selector=app.kubernetes.io/instance=owg
    NAME     READY     STATUS      RESTARTS     AGE
    owg-opentelemetry-collector-594fddd656-
    tfstk    1/1       Terminating   1 (70s ago)    2m8s
    owg-opentelemetry-collector-7955d689c4-
    gsvqm    1/1       Running       0              3s
    ```

You will now have a lot more labels available for your Loki log data. Let's explore what that means in the next section.

Introducing Loki

Grafana Loki was designed from the ground up to be a highly scalable multi-tenant logging solution. Its design was heavily influenced by Prometheus with a few main objectives:

- It was built with developers and operators in mind (such as *Diego* and *Ophelia*, who were introduced in *Chapter 1*)

- It has simple ingestion; no pre-parsing is required

- It only indexes metadata about logs

- It stores everything in an object store

Let's look at how Loki ingests data and uses labels as this will provide valuable insight into the way your queries source and then process the data for presentation:

- **Log ingest**: Loki accepts logs from all sources with a wide choice of agents available to make that easy. You can even send log data directly to the Loki API. This makes it the perfect choice for complex environments featuring a multitude of systems and hardware components.

  Loki stores its logs as log streams, where each entry has the following:

  - **Timestamp**: It has nanosecond precision for accuracy.

  - **Labels**: These are key-value pairs used for the identification and retrieval of your data; they form the Loki index.

  - **Content**: This refers to the raw log line. It is not indexed and is stored in compressed chunks.

  The following diagram shows a log stream with a log line and its associated metadata:

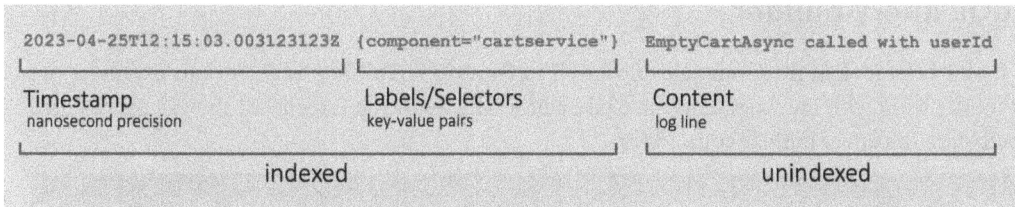

Figure 4.1 – Loki log structure

- **Log labels**: Loki log labels provide the metadata for the log line and not only help identify the data but also are used to create the index for the log streams and structure the log storage. They have the following features:

  - Each unique set of labels and values creates a log stream

  - Logs in a stream are batched, compressed, and stored as chunks

  - Labels are the index to Loki's log streams

  - Labels are used to search for logs

The following diagram demonstrates two log streams. As you can see, in a stream of logs, each log has the same unique set of labels. In this instance, `k8s_node_name` has two values:

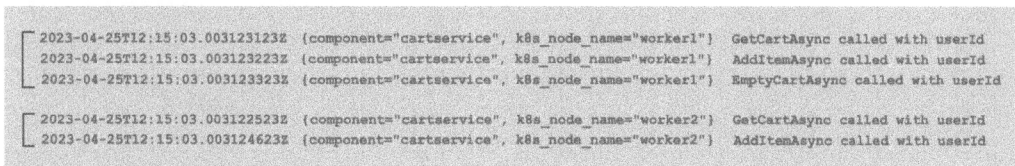

Figure 4.2 – Loki log streams

Now that we have looked at the structure of a Loki log, let's introduce **LogQL**, the query language used to extract value from your logs.

Understanding LogQL

Grafana developed LogQL as the query language for Loki using the **Prometheus Query Language** (**PromQL**) for inspiration. It was designed with developers (*Diego*) and operators (*Ophelia*) in mind (you can refer to *Chapter 1* for an introduction to these personas), providing familiar filtering and aggregation mechanisms. Loki does not index the log content. Log events are grouped into log streams and indexed with labels (the log metadata). Executing a LogQL query in Loki invokes a type of distributed filtering against log streams to aggregate the log data.

Let's explore the **Grafana explorer** UI for LogQL, where you will be executing most of your LogQL queries.

LogQL query builder

We took a brief look at the Grafana explorer UI in *Figure 3.16* in *Chapter 3*. For our examples, we will mostly work with raw LogQL in the **Code** editor. The following screenshot shows LogQL typed directly into the query builder code editor:

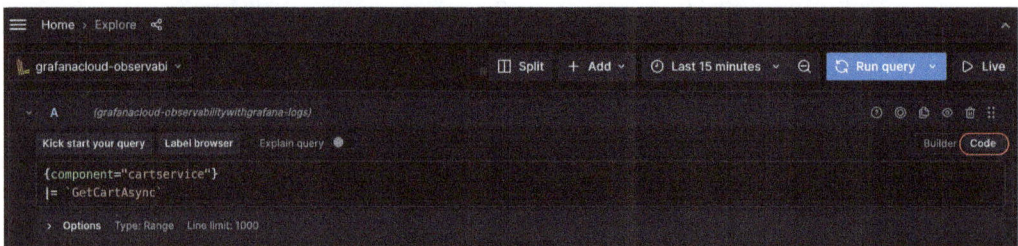

Figure 4.3 – LogQL query builder Code editor

If you ever get stuck with your LogQL, you can lean on the **Log query starters** and **Explain query** tools to help you get started with your queries and understand what each step of your pipeline is doing.

Log query starters provides some quick examples to work with your data and get you filtering and formatting it with ease:

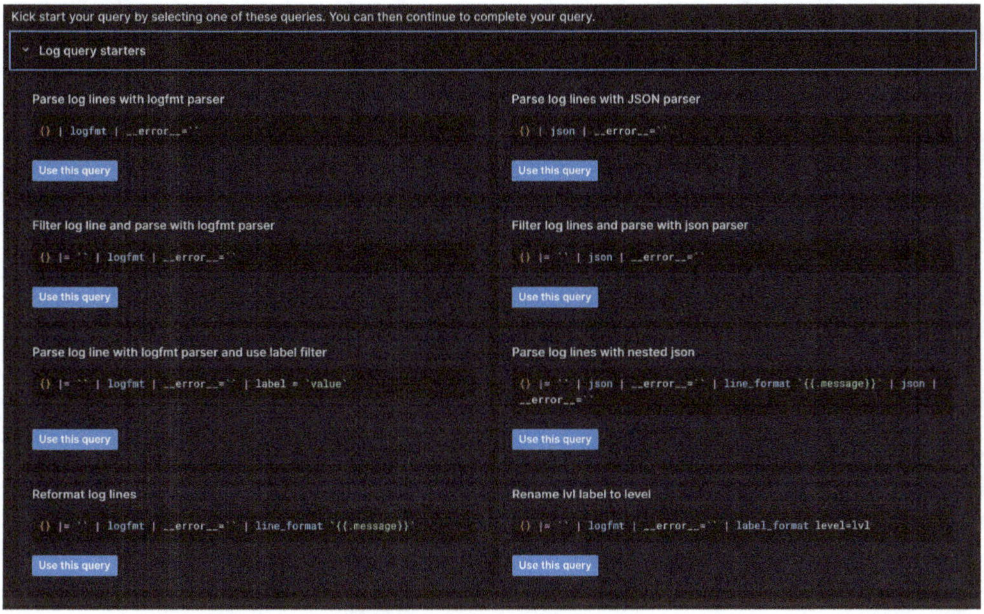

Figure 4.4 – Log query starters

Similarly, **Metric query starters** provides some quick examples to work with your data and generate metrics ready for use in dashboards and alerts:

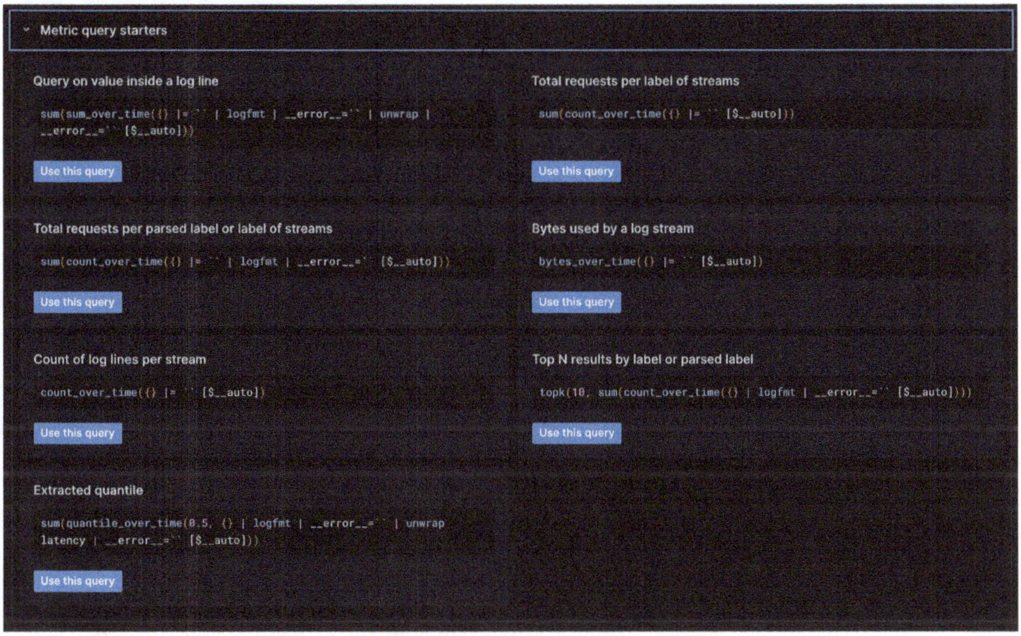

Figure 4.5 – Metric query starters

Available in the LogQL query builder and the dashboard panel editor, **Explain query**, when toggled on, provides a breakdown of each stage of your LogQL pipeline. This tool is invaluable when analyzing an existing query or debugging your own during design:

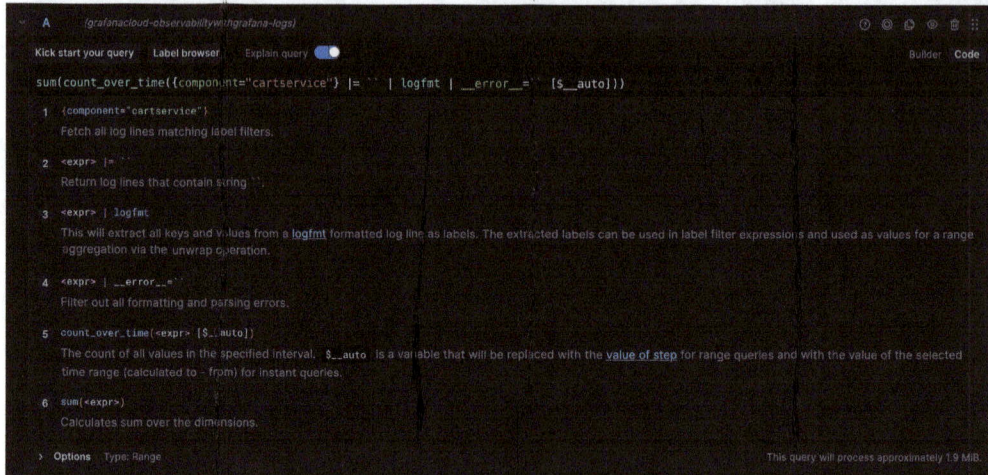

Figure 4.6 – Explain query

Let's now explore the features of LogQL available for selecting, filtering, and parsing your log data.

An overview of LogQL features

A basic LogQL query consists of one or more log stream selectors to retrieve the raw log chunks for processing and an optional log pipeline to filter and parse the log data. The following figure shows a basic LogQL query with the `component="cartservice"` selector and a pipeline filter, `|= `GetCartAsync``, which would return two lines from the log stream example in *Figure 4.2*:

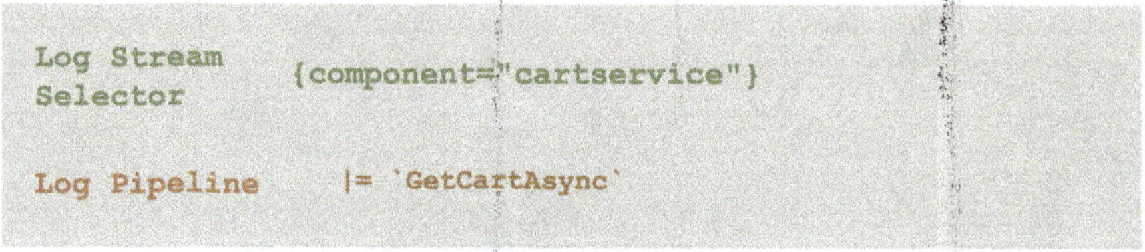

Figure 4.7 – A basic LogQL query

The following reference table shows the different features available to you when building your LogQL query, which will help while you get familiar with querying your logs with Loki:

LogQL Sections	Syntax	Operators	Scope
Stream selector	`{label="value", foo!="bar"}`	`=,` `!=,` `=~,` `!~`	Select log streams to retrieve; there must always be at least one selector
Line filter	`|= \`error\``	`|=,` `!=,` `|~,` `!~`	Filter to matching log lines
Parser	`| json`	`json, logfmt, pattern, regexp, unpack`	Parse and extract labels from the log content, with parses for structured and unstructured logs
Label filter	`| label="value"`	`=,` `!=,` `=~,` `!~,` `<,` `<=,` `>,` `>=`	Filter log lines using original and newly extracted labels
Line format	`| line_format "{{. label}}"`		Rewrite the log line content for presentation purposes
Label format	`| new_label="{{. label}}"`		Rename, modify, or add labels

Table 4.1 – LogQL feature overview

Let's start by looking at the log stream selector in detail.

Log stream selector

Selecting log streams to include in your query results requires filtering based on the Loki labels using simple operators. By improving the granularity of your log stream selector, you can reduce the number of streams searched and improve query performance. We will discuss this in more detail later in this chapter when we look at the Loki architecture. There must always be at least one stream selector, but it must be written in a way that will not match empty values; for example, the **regular expression** (**regex**) `{label=~".*"}` will fail as a 0 or more quantifier and `{label=~".+"}` will pass as a 1 or more quantifier. Multiple stream selectors can be used and are separated using commas. The log stream selector must be the first item in your LogQL query and identified with curly braces. For example, the following LogQL query will select log streams where the `component` label ends with `service` and the `name` label equals `owg-demo-checkoutservice`:

```
{component=~".+service", name="owg-demo-checkoutservice"}
```

As mentioned at the beginning of this section, LogQL was inspired by PromQL and as such, the Prometheus label selector rules have been adopted by LogQL for log stream selectors:

Operator	Meaning
=	Exactly equal
!=	Not equal
=~	Regex matches
!~	Regex does not match

Table 4.2 – Log stream selector operators

Grafana has implemented the **Golang RE2 syntax** for log streams, which means you will have to match against entire strings. This includes newlines, so it's worth checking this if your regex filters are failing. Syntax documentation can be found here: `https://github.com/google/re2/wiki/Syntax`.

Once you have your log streams selected, the **log pipeline** can then be used to filter and process them. Let's discuss this next.

Log pipeline

As we presented in *Table 4.1*, the expressions available are line and label filters, parsers, and formatters. Expressions are executed in sequence for each line of the log stream, dropping anything filtered out and moving on to the next line.

Expressions allow you to transform or mutate the log data to then use it for additional filtering/processing. Let's look at the following example (this is the pipeline section only; you would need the `{component=~".+service"}` selector to make it work in Grafana):

```
|= `emailservice`
| json
| resources_k8s_container_restart_count > 0
| line_format `{{.body}}`
| __error__=``
```

Here, we're doing the following tasks:

1. We start by matching the logs containing `emailservice`.

2. We then use the `json` parser to extract additional labels that are filtered where `resources_k8s_container_restart_count` is greater than 0.

3. We then rewrite the log line to only contain the contents of `body`.

4. Finally, we strip all formatting and parsing errors.

Let's now look at each of the log pipeline expressions and how to use them.

Line filters

Grafana describes **line filters** as a distributed `grep` over the aggregated logs from the matching log streams. We will understand this statement better in the *Loki's architecture* section. For now, it's fine to just understand line filters as case-sensitive searches through log line contents dropping lines that do not match. Filter expressions are made up of a filter operator followed by text or regex. The following table shows the meaning of each expression with examples:

Operator	Meaning	Example
`|=`	Log line contains a string	`|= `emailservice``
`!=`	Log line does not contain a string	`!= `emailservice``
`|~`	Log line contains a match to the regex	`|~ `email\w+``
`!~`	Log line does not contain a match to the regex	`!~ `email\w+``

Table 4.3 – Line filters

It is best practice to start log pipelines with line filter expressions to reduce the result set for subsequent expressions and improve the performance of the query.

IP address matching

LogQL provides a useful function to aid **IP address matching** without complex regex patterns. Using the ip("<pattern>") syntax, it supports both IPv4 and IPv6 addresses, address ranges, and CIDR patterns. This function works for both line and label filters with a slight caveat in implementation; only | = and ! = are allowed for line filter expressions. We'll look at this in the context of label filters later in this section.

The following examples show various patterns (ip(<pattern>)) along with an explanation of what each will do, for both IPv4 and IPv6:

- **Match a single IP address**: ip("192.168.0.22") and ip("::1")
- **Find a match within a range**: ip("192.168.0.1-192.189.10.12") and ip("2001:db8::1-2001:db8::8")
- **Find a match within a CIDR specification**: ip("192.52.100.0/24") and ip("2001:db8::/32")

Decolorize

In *Chapter 2*, we mentioned unstructured logging is often color-coded to improve readability on the computer terminal. However, in a log aggregation system, such as Loki, those color codes are displayed in full. For example, the color red would be displayed as \u001b[31m.

Loki has a simple line filter expression that removes these ANSI sequences for color codes so that you can clean the log to make it more readable in Grafana:

```
{name="emailservice"} | decolorize
```

Parsers

We have said this before: Loki accepts logs from all sources. It does not really matter what your logs look like; they can come in structured, semi-structured, or unstructured formats. It is, however, important when designing and building observability solutions to understand the log formats you are working with. This ensures that you can ingest, store, and parse log data in a way it can be used effectively. The personas in *Chapter 1* give you an idea of who these will be used by and for what purpose.

Having a good understanding of your source log format is important to instruct you on what to use and help with your overall observability design. The following LogQL parsers can be used to parse and extract labels from your log content:

- json: If your log content is structured or semi-structured JSON and has embedded JSON (which can be isolated using the line_format expression), the **JSON parser** can be used. Using | json on its own will extract all of the JSON properties as labels. Any nested properties will be represented as a single label separated with _. Arrays are skipped completely when extracting all of the properties. Additionally, expressions can be passed into the JSON parser as quoted strings to restrict the output to only the labels required, for example, | `json label1="expression", label2="expression"`, where the expression identifies a key or nested key. Arrays are returned where identified by expressions and they are assigned to the label formatted as JSON.

- `logfmt`: If your log content is structured, single-level key-value pairs, it can be parsed with the **logfmt parser**. Using | `logfmt` on its own will extract all of the key-value pairs. Similar to the JSON parser, expressions can be passed into the `logfmt` parser as quoted strings to restrict the output to only the labels required, for example, | `logfmt label1="expression", label2="expression"`, where `expression` identifies a key.

- `pattern`: For unstructured log content, the **pattern parser** allows the explicit extraction of fields from log lines using | `pattern "<expression>"`, where `expression` matches the structure of a log line. The pattern parser expression is made up of captures delimited by the < and > characters and literals, which can be any sequence of UTF-8 characters.

- `regexp`: Unstructured log content can also be extracted using the **regular expression parser**, which takes a single expression, | `regexp "<expression>"`, where `expression` is a regex pattern that complies with the Golang RE2 syntax. A valid expression must contain at least one sub-match, with each sub-match extracting a different label.

- `unpack`: If you are using a compatible logging agent, such as Grafana Agent or Promtail, you can take advantage of the `unpack` parser to unpack embedded labels created by Promtail's `pack` feature. With Promtail's `pack` feature, the original log line is stored in the `_entry` key. This value will be used to replace the log line.

Label filters

We discussed log labels at the beginning of this section with regard to log ingestion and retrieval. Additionally, labels can be extracted as part of the log pipeline using parser and formatter expressions. The **label filter** expression can then be used to filter your log line with either of these labels.

The statement part of the filter is referred to as the *predicate*, and in the case of label filters, it contains the following:

- The label identifier
- The operation
- A value

For example, in `name="emailservice"`, the `name` label is compared, using the = operator, with the value `"emailservice"`. It is processed from left to right, so the label identifier must start the predicate.

Value types are inferred from your query input. In the following table, you will find an overview of these types as a useful reference for building your label filters:

Value Type	Description
String	Can be surrounded with double quotes or backticks, for example, `"emailservice"` or `` `emailservice` ``.
Duration	Structured as a sequence of decimal numbers. They can optionally contain fractions and the unit can be declared as a suffix, for example, `"280ms"`, `"1.3h"`, or `"2h30m"`. Units of time that can be used are `"ns"`, `"us"` (or `"µs"`), `"ms"`, `"s"`, `"m"`, and `"h"`, and as the examples show, you can use multiple units: `"2h30m"`.
Number	Standard floating-point numbers, for example, `357` or `98.421`.
Bytes	Structured as a sequence of decimal numbers. They can optionally contain fractions and the unit can be declared as a suffix, for example, `"36MB"`, `"2.4Kib"`, or `"18b"`. Units for bytes that can be used are `"b"`, `"kib"`, `"kb"`, `"mib"`, `"mb"`, `"gib"`, `"gb"`, `"tib"`, `"tb"`, `"pib"`, `"pb"`, `"eib"`, and `"eb"`.

Table 4.4 – Value types

Let's now look at these value types in more detail:

- **String**: As with the label matchers we use with the log stream selector, the =, !=, =~, and !~ operations can be used. The `string` type is used to filter the built in label __error__, which is often used to strip formatting and parsing errors from results; for example, `| __error__=``.

- **Duration, Number, and Bytes**: All of the remaining value types follow the same rules. Duration, Number, and Bytes convert the label value for use with the following list of comparators:

Operator	Meaning
= or ==	Equals
!=	Does not equal
> and >=	Is greater than or greater than and equal to
< and <=	Is less than or less than and equal to

Table 4.5 – Type operators

Take the example `| resources_k8s_container_restart_count > 0`. Loki attempts to convert the value for use with the operator if it needs to. If there are any errors with the conversion, the __error__ label will be added to the log line, which as we demonstrated earlier, can be filtered out using `| __error__=``.

Grafana LogQL also allows for multiple predicates to be *chained* together using and and or . and can alternatively be expressed using , or | or <space>.

For example, all of the following produce the same output:

```
| quantity >= 2 and productId!~"OLJ.*"
| quantity >= 2 | productId!~"OLJ.*"
| quantity >= 2 , productId!~"OLJ.*"
| quantity >= 2 productId!~"OLJ.*"
```

We described **IP address matching** in the *Line filters* section. **Label filter expressions** are the same except only the = and ! = label matchers are allowed. Once you have filtered and parsed your logs as required, you can begin to transform the data, whether that is for presentation or further pipeline processing. We will discuss the two ways of doing this in detail next. But first, let's explore **template functions**, which are implemented by both line and label filters.

Template functions

The Golang text/template format has a large set of available template functions, all of which are available for use in LogQL queries. Full documentation can be found on the Grafana website. The **templating engine** has access to your data in various ways:

- It can treat labels as variables, referencing them using . , for example, {{ .component }}

- It can access the log line itself using __line__, for example, `{{ __line__ | lower }}`

- It can access the log timestamp using __timestamp__, for example, `{{ __timestamp__ | date "2023-04-25T12:15:03.00Z+01:00" }}`

Template functions can be broken down into the following distinct areas:

- Regex patterns

- String functions

- Math functions

- JSON functions

- Date and time functions

In addition, there are other functions that do not necessarily fit into a grouping but are nevertheless very useful. These include encode and decode functions, byte and duration conversions, counts, and default values.

Line and label format

Two features are available for transforming logs:

- **Line format**: The line format expression, given by `| line_format "{{ .label }}"`, is used to rewrite log line content. This expression is used to modify your log line using the template functions referenced earlier. LogQL injects all labels as variables into the template, making them available for use, for example, `| line_format "{{.label_one}} {{.label_two}}"`. The format takes double quotes or backticks, where backticks allow you to avoid escaping characters.

  For example, if we have the following labels, `method=sent`, `status=200`, and `duration=15ms`, the following LogQL query would return `sent 200 15ms`:

  ```
  {instance="owg-demo", component="featureflagservice"} |= `Sent`
  | json
  | regexp "(?P<method>Sent) (?P<status>\\d+?)
  in\\s(?P<duration>.*?ms)"
  | line_format "{{.method}} {{.status}} {{.duration}}"
  ```

- **Label format**: The label format expression given by `|label_format new_label="{{ .label }}"` is used to rename, modify, or even create new labels. It accepts a comma-separated list of equality operations, allowing multiple operations to be carried out simultaneously.

  To rename a label with another label, the label identifiers must be on both sides of the operator; for example, `target=source` will put the contents of the `source` label into the `target` label and drop the `source` label.

  A label can be populated using the Golang text/template format and functions detailed previously (double quotes or backticks). For example, if we have the `user=diego` and `status=200` labels, the `|label_format target="{{.status}} {{.user}}"` pipeline would define the `target` label as `200 diego`.

  Templating can be used if you wish to preserve the original `source` label. In the example, `target=source` removed the `source` label. We can write this as `target="{{.source}}"`, which will put the contents of the `source` label into the `target` label while preserving the `source` label. If the target label does not already exist, a new label is created.

> **Important note**
>
> Only one instance of a label name can be used per expression; for example, `| label_format foo=bar,foo="new"` would fail. The desired result could be implemented with two expressions, one following the other, like this: `| label_format foo=bar | label_format foo="new"`.

We've looked at how the label format gives you options to create, modify, and rename labels. Additionally, we have the `drop labels` command to remove labels completely. Let's explore that expression now.

Dropping labels

The `drop labels` expression is used to remove labels from the pipeline. For example, if we have the `user=diego`, `status=200`, and `duration=1000 (ms)` labels, the `|drop user` pipeline would drop the `user` label, leaving only `status` and `duration`.

We will now take a look at more of the LogQL features, exploring formatters, metric queries, and the UI for executing LogQL the **Grafana Explorer** where queries for all data sources are built.

Exploring LogQL metric queries

One of the most powerful features of Loki and LogQL is the ability to create metrics from logs. With **metric queries**, you can, for example, calculate the rate of errors or the top 10 log sources with the highest volume of logs over the last hour. This makes it perfect for creating visualizations or triggering alerts.

If we combine metric queries with the parsers and formatters we looked at earlier in this section, they can be used to calculate metrics from sample data within a log line. For example, latency or request size can be extracted from log data and used as a metric. These will then be available for aggregations and the generation of new series.

Let's now take a look at the aggregations available, namely, **range vector aggregations** and **built-in aggregation operators**.

Range vector aggregations

The Prometheus concept of a **range vector** is shared by LogQL, where the range of samples is a range of log or label values. We will discuss the range vector concept in greater detail in *Chapter 5*. The selected aggregation is applied to a time interval specified as a number followed by a unit. The following time interval units can be used:

- `ms`: Milliseconds
- `s`: Seconds
- `m`: Minutes
- `h`: Hours
- `d`: Days
- `w`: Weeks
- `y`: Years

Examples are `6h`, `1h30m`, `10m`, and `20s`.

There are two types of range vector aggregations supported by Loki and LogQL: **log range aggregations** and **unwrapped range aggregations**. Let's explore these in detail.

Log range aggregation

A **log range aggregation** is a LogQL query followed by a duration, for example, `[10ms]`, with a function applied to it to aggregate the query over the duration. The duration can be placed after the log stream selector or at the end of the log pipeline.

Here are the aggregation functions:

Aggregation	Description
`rate(range)`	Will calculate the number of entries per second.
`count_over_time(range)`	Will count the entries for each log stream within the given range.
`bytes_rate(range)`	Useful to detect changes in log data volume. It will calculate the number of bytes per second for each log stream.
`bytes_over_time(range)`	Useful to calculate the volume of log data. It will count the amount of bytes used by each log stream for the given range.
`absent_over_time(range)`	Useful for alerting when there are no time series and logs streams for label combinations for a duration of time. It returns an empty vector if the range passed to it has elements and a single element vector with the value 1 if the range passed to it has no elements.

Table 4.6 – Log range aggregation functions

Here are a few log range aggregation examples:

- To count all the log lines within the last 10 minutes for the `currencyservice` component:

  ```
  count_over_time({component="currencyservice"}[10m])
  ```

- To sum the rate per second of errors by component within the last minute:

  ```
  sum by (component) (rate({component=~".+service"}
  |= "error" [1m]))
  ```

Unwrapped range aggregations

Unwrapped range aggregations use the LogQL unwrap function to extract a value to be used in the aggregation. They support grouping using the by or without clause to aggregate over distinct labels. The without aggregation removes the labels identified from the result vector while preserving all other labels. The by aggregation drops labels that are not identified in the by clause.

Here are the aggregation functions:

Aggregation	Description
`rate(unwrapped-range)`	Will calculate the per-second rate of the sum of all of the values within the interval.
`rate_counter(unwrapped-range)`	Will calculate the per-second rate of all the values within the interval, treating them as counter metrics.
`sum_over_time(unwrapped-range)`	Will sum of all the values within the interval.
`avg_over_time(unwrapped-range)`	Will return the average of all the points within the interval.
`max_over_time(range)`	Will return the maximum of all the points within the interval.
`min_over_time(unwrapped-range)`	Will return the minimum of all the points within the interval.
`first_over_time(unwrapped-range):`	Will return the first value of all the points within the interval.
`last_over_time(unwrapped-range)`	Will return the last value of all the points within the interval.
`stdvar_over_time(unwrapped-range)`	Will return the population standard variance of the values within the interval.
`stddev_over_time(unwrapped-range)`	Will return the population standard deviation of the values within the interval.
`quantile_over_time(scalar, unwrapped-range)`	Will return the specified quantile of the values within the interval.
`absent_over_time(unwrapped-range)`	Useful for alerting when there are no time series and logs streams for label combinations for a duration of time. It returns an empty vector if the range passed to it has elements and a single element vector with the value 1 if the range passed to it has no elements.

Table 4.7 – Unwrapped range aggregation functions

The `sum_over_time`, `absent_over_time`, `rate`, and `rate_counter` functions are excluded from grouping.

Here are a few unwrapped range aggregation examples:

- To calculate the 99th percentile of the `webserver` container `request_time` excluding any JSON formatting errors by `path` within the last minute:

```
quantile_over_time(0.99,
  {container="webserver"}
    | json
    | __error__ = ""
    | unwrap duration_seconds(request_time) [1m]) by (path)
```

- To calculate the number of bytes processed by `org_id` within the last minute, filtering where the log contains the `metrics` string:

```
sum by (org_id) (
  sum_over_time(
  {container="webserver"}
    |= "metrics"
    | logfmt
    | unwrap bytes(bytes_processed) [1m])
  )
```

Built-in aggregation operators

LogQL supports a subset of the **built-in aggregation operators** that PromQL supports. These can be used to aggregate the element of a single vector, resulting in a new vector of fewer elements but with aggregated values.

The following table shows some built-in range aggregation operators:

Aggregation	Description
sum	Will calculate the sum by the labels specified
avg	Will calculate the average by the labels specified
min	Will select the minimum by the labels specified
max	Will select the maximum by the labels specified
stddev	Will calculate the population standard deviation by the labels specified
stdvar	Will calculate the population standard variance by the labels specified
count	Will count the number of elements in a vector
topk	Will select the largest k elements by sample value
bottomk	Will select the smallest k elements by sample value

sort	Will return the vector elements sorted by their sample values, in ascending order
sort_desc	Will return the vector elements sorted by their sample values, in descending order

Table 4.8 – Built-in range aggregation functions

Here are a few built-in range aggregation operator examples:

- To return the top 10 applications by the highest log throughput for the last 10 minutes by `name`:

```
topk(10, sum(rate({region="us-west1"}[10m])) by (name))
```

- To return the average rate of `GET` requests to the `/hello` endpoint for web server logs by region for the last 10 seconds:

```
avg(rate(({container="webserver"} |= "GET" | json | path="/
hello")[10s])) by (region)
```

We have looked at how LogQL can parse different log formats. Let's now take a look at the Loki architecture and how Loki stores and queries the log data you send.

Exploring Loki's architecture

Grafana Loki has a full **microservices architecture** that can be run as a single binary and a simple scalable deployment to a full microservices deployment running all the components as distinct processes. At a high level, it is made up of features that implement write, read, and store functionality, as shown in the following diagram:

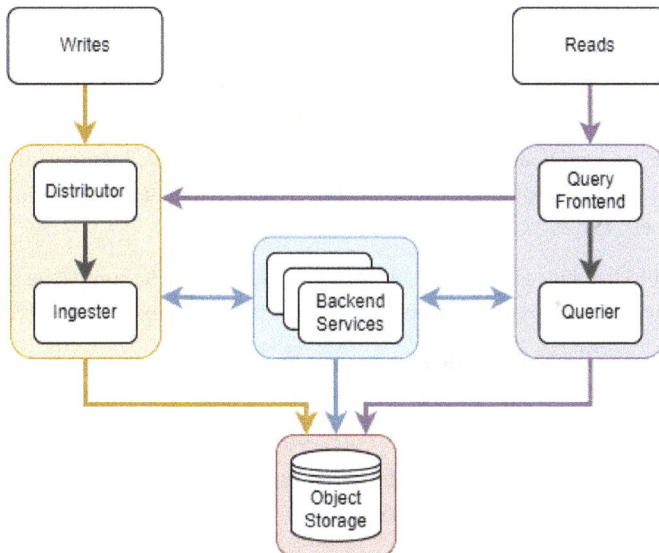

Figure 4.8 – High-level overview of Loki architecture

Both *write* and *read* functionality can be scaled independently to suit your particular needs and use cases.

Under the hood, Loki has the following core components:

- Distributor
- Ingester
- Query frontend
- Querier
- Backend services:
  - Ruler
  - Compactor
  - Query scheduler

Let's now look at the functionality and core components in more detail:

- **Writes**: Writes for the incoming log data hit the distributor, which is responsible for data sharding and partitioning, and sending them to the ingesters. The distributor validates each set of streams, checking labels, timestamps, and log line sizes, then batches log stream chunks to multiple ingesters.

  The ingester writes to the **write-ahead logs** (**WALs**) for resiliency and finally into the object storage backend.

  Both the querier and ruler read the ingester to access the most recent data. The querier can additionally access the object storage data.

- **Reads**: The query frontend is responsible for accelerating query execution, distributing large queries across multiple queriers and ensuring retries in the event of failure.

  Queriers parse the LogQL and query the underlying systems: the ingester for the most recent data and object storage for older data. The querier de-duplicates data with the same nanosecond timestamp, labels, and log content.

- **Storage**: The object storage is where the batched logs are stored. The compactor is responsible for maintaining the data. It monitors the object storage, de-duplicating data and removing old logs.

- **Backend services**: The ruler evaluates queries and performs actions based on the result. The actions can be recording rules (generating new metrics for LogQL queries) or alerts for system events.

  The alert manager is responsible for notifications and alerts triggering and being sent from the system, but this is not included with Loki.

- **Loki index**: In this chapter, so far, we have covered Loki log labels and LogQL log stream selectors. The underlying architecture completes the picture, explaining how the distributor shards the data. It is that sharding and subsequent storage using the label-based Loki index that makes Loki fast and inexpensive. It also validates the importance of a good labeling strategy to improving storage and retrieval, and essentially querying performance.

Now that we have built up a good understanding of Loki, let's look at a few best practices and some tips for working with Loki log data.

Tips, tricks, and best practices

In this section, we will look at a few best practices for filtering and cardinality. We will then look at the LogQL Analyzer and LogCLI, which are tools that can help you when you are working with Grafana Loki log data.

Here are some best practices to keep in mind:

- **Filter first**: Loki stores the raw log in object storage as compressed chunks. Because of this, it is important, from a speed point of view, to filter early. Processing complex parsing on smaller datasets will increase the response time.

- **Cardinality**: High cardinality in Loki can be very detrimental. It is important to design your Loki labels well. Anything that has a lot of variable data in it is a bad idea as that will multiply the number of log streams and therefore storage chunks by that factor. Thinking of them as a locator rather than a content descriptor helps. You can always extract labels from log lines with the range of parsers available. Some examples of good labels (targeted with limited values) are the following:

  - `namespace`

  - `cluster`

  - `job`

  - `app`

  - `instance`

  - `filename`

  Some examples of poor labels (often vague with unlimited values) are the following:

  - `userid`

  - `traceid`

  - `path`

  - `status code`

  - `date`

Now, let's take a closer look at the advantages offered by using the LogQL Analyzer and LogCLI:

- **LogQL Analyzer**: The LogQL Analyzer provides an interface on the Grafana website for you to practice your LogQL queries. You can view detailed explanations of the actions implemented by your query on a sample log entry. Head over to `https://grafana.com/docs/loki/latest/query/analyzer/` to try it out. Let's take a look at the Loki LogQL Analyzer:

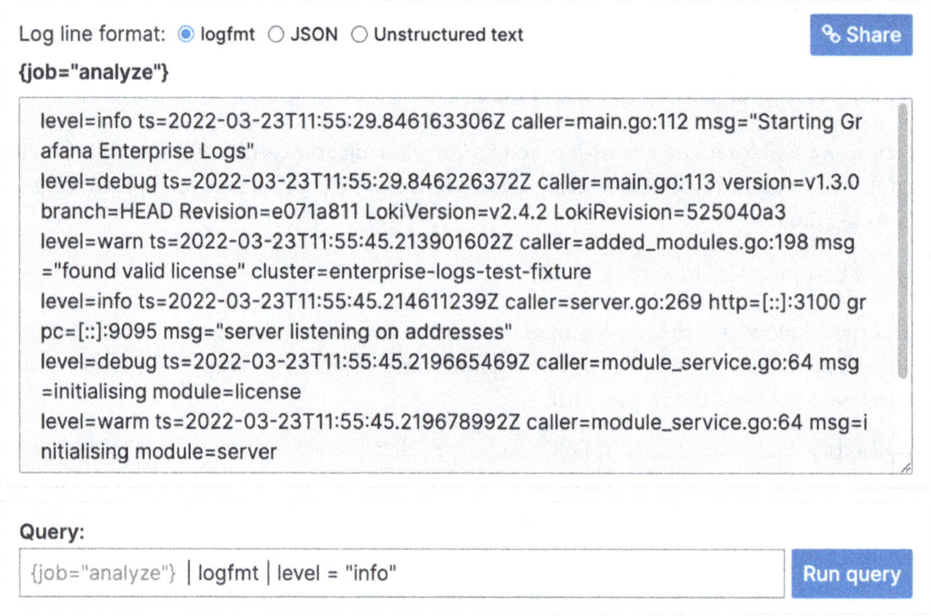

Log line format: ◉ logfmt ○ JSON ○ Unstructured text   % Share

{job="analyze"}

```
level=info ts=2022-03-23T11:55:29.846163306Z caller=main.go:112 msg="Starting Gr
afana Enterprise Logs"
level=debug ts=2022-03-23T11:55:29.846226372Z caller=main.go:113 version=v1.3.0
branch=HEAD Revision=e071a811 LokiVersion=v2.4.2 LokiRevision=525040a3
level=warn ts=2022-03-23T11:55:45.213901602Z caller=added_modules.go:198 msg
="found valid license" cluster=enterprise-logs-test-fixture
level=info ts=2022-03-23T11:55:45.214611239Z caller=server.go:269 http=[::]:3100 gr
pc=[::]:9095 msg="server listening on addresses"
level=debug ts=2022-03-23T11:55:45.219665469Z caller=module_service.go:64 msg
=initialising module=license
level=warm ts=2022-03-23T11:55:45.219678992Z caller=module_service.go:64 msg=i
nitialising module=server
```

Query:

{job="analyze"} | logfmt | level = "info"   Run query

Figure 4.9 – Loki LogQL Analyzer

The explanations provided by the LogQL Analyzer are far more detailed than the **Explain query** feature in the query builder, so it's worth checking out while you are learning LogQL.

- **Using LogCLI**: For command-line lovers everywhere, Grafana Loki comes with a command-line interface called LogCLI that allows you to do the following at your terminal:

  - Query your logs

  - Evaluate metric queries for a single point in time

  - Identify Loki labels and obtain stats about their values

  - Return log streams for a time window with a label matcher

This is great if you need access to the power of LogQL without leaving the comfort of your own console.

Full setup documentation and command references can be found here: `https://grafana.com/docs/loki/latest/query/`. You can download the binary from the Loki releases page on GitHub.

We will now wrap up this chapter with a reminder of what you have learned.

Summary

In this chapter, we have taken a look at Loki, exploring the log ingest format and the importance of log labels. We then started looking at the comprehensive features of LogQL, the query language of Loki, and how we can select log streams and then filter, parse, format, and transform log lines. These techniques will be invaluable when working with Loki to build dashboards in *Chapter 8*. Then, we looked at the Loki architecture to get an understanding of what's going on behind the scenes. We also explained how our data is stored and how Loki can be scaled to increase performance. Lastly, we reviewed some tips and best practices that can help you improve your experience with Loki.

In the next chapter, we'll move on from logs to explore **metrics** and **Prometheus**, where Loki took its original inspiration from.

5

Monitoring with Metrics Using Grafana Mimir and Prometheus

This chapter will introduce the **Prometheus query language** (**PromQL**). Like LogQL, PromQL can be used to select and filter metrics streams and process numeric data with operators and functions, enabling you to build quick and efficient queries that will support establishing an observable system. We will also explore and compare the various protocols that can be used to output metrics from systems. Finally, we will explore the architecture of **Prometheus** and **Mimir** to understand how Mimir fills the need for a highly scalable system.

We will cover the following main topics in this chapter:

- Updating the OpenTelemetry collector for metrics
- Introducing PromQL
- Exploring data collection and metric protocols
- Understanding data storage architectures
- Using exemplars in Grafana

Technical requirements

In this chapter, you will need the following:

- The OpenTelemetry demo application set up in *Chapter 3*
- The Grafana Cloud instance set up in *Chapter 3*
- Docker and Kubernetes

You'll find the code for this chapter in the GitHub repository at `https://github.com/PacktPublishing/Observability-with-Grafana/tree/main/chapter5`. You'll find the *Code in Action* videos for this chapter at `https://packt.link/A2g91`.

Updating the OpenTelemetry demo application

For this chapter, we have prepared an updated version of `OTEL-Collector.yaml`, which will add additional labels to metrics for you to explore. Full details on this process are available from the Git repository in the `README.md` file. This process will apply the new version of the collector configuration to your demo application:

1. Using Helm, we will apply the updated configuration file to our Kubernetes cluster:

```
$ helm upgrade --version '0.73.1' --values chapter5/OTEL-
Collector.yaml --values OTEL-Creds.yaml owg open-telemetry/
opentelemetry-collector
NAME: owg-otel-collector
LAST DEPLOYED: Sun Mon 19 12:42:36 2023
NAMESPACE: default
STATUS: deployed
REVISION: 2
...
```

2. Validate upgrade was successful:

```
$ kubectl get pods --selector=component=standalone-collector
NAME   READY   STATUS   RESTARTS   AGE
owg-otel-collector-594fddd656-tfstk   1/1   Terminating   1
(70s ago)   2m8s
owg-otel-collector-7b7fb876bd-
vxgwg   1/1   Running   0   3s
```

This new configuration adds the collection of metrics from the Kubernetes cluster and the OpenTelemetry collector. The configuration also does some necessary relabeling.

Now that we are collecting more data from our local demo application, let's introduce the language used to query that data.

Introducing PromQL

Prometheus was initially developed by SoundCloud in 2012; the project was accepted by the *Cloud Native Computing Foundation* in 2016 as the second incubated project (after Kubernetes), and version 1.0 was released shortly after. PromQL is an integral part of Prometheus, which is used to query stored data and produce dashboards and alerts.

Before we delve into the details of the language, let's briefly look at the following ways in which Prometheus-compatible systems interact with metrics data:

- **Ingesting metrics**: Prometheus-compatible systems accept a timestamp, key-value labels, and a sample value. As the details of the **Prometheus Time Series Database** (**TSDB**) are quite complicated, the following diagram shows a simplified example of how an individual sample for a metric is stored once it has been ingested:

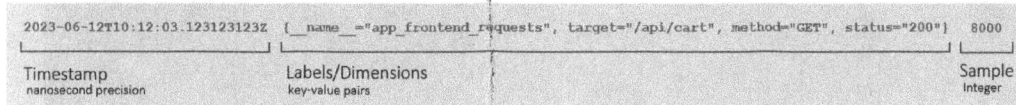

```
2023-06-12T10:12:03.1231231232Z   {__name__="app_frontend_requests", target="/api/cart", method="GET", status="200"}   8000

Timestamp                          Labels/Dimensions                                                                    Sample
nanosecond precision               key-value pairs                                                                      Integer
```

Figure 5.1 – A simplified view of metric data stored in the TSDB

- **The labels or dimensions of a metric**: Prometheus labels provide metadata to identify data of interest. These labels create metrics, time series, and samples:

  - Each unique __name__ value creates a **metric**. In the preceding figure, the metric is app_frontend_requests.

  - Each unique set of labels creates a **time series**. In the preceding figure, the set of all labels is the time series.

  - A time series will contain multiple **samples**, each with a unique timestamp. The preceding figure shows a single sample, but over time, multiple samples will be collected for each time series.

  - The number of unique values for a metric label is referred to as the **cardinality** of the label. Highly cardinal labels should be avoided, as they significantly increase the storage costs of the metric.

  The following diagram shows a single metric containing two time series and five samples:

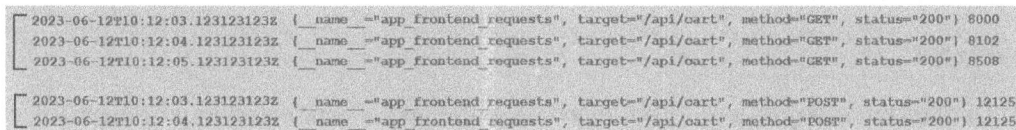

```
2023-06-12T10:12:03.1231231232Z   {__name__="app_frontend_requests", target="/api/cart", method="GET", status="200"}   8000
2023-06-12T10:12:04.1231231232Z   {__name__="app_frontend_requests", target="/api/cart", method="GET", status="200"}   8102
2023-06-12T10:12:05.1231231232Z   {__name__="app_frontend_requests", target="/api/cart", method="GET", status="200"}   8508

2023-06-12T10:12:03.1231231232Z   {__name__="app_frontend_requests", target="/api/cart", method="POST", status="200"}  12125
2023-06-12T10:12:04.1231231232Z   {__name__="app_frontend_requests", target="/api/cart", method="POST", status="200"}  12125
```

Figure 5.2 – An example of samples from multiple time series

In Grafana, we can see a representation of the time series and samples from a metric. To do this, follow these steps:

1. In your Grafana instance, select **Explore** in the menu.

2. Choose your Prometheus data source, which will be labeled as grafanacloud-<team>-prom (default).

3. In the **Metric** dropdown, choose **app_frontend_requests_total**, and under **Options**, set **Format** to **Table**, and then click on **Run query**. This will show you all the samples and time series in the metric over the selected time range. You should see data like this:

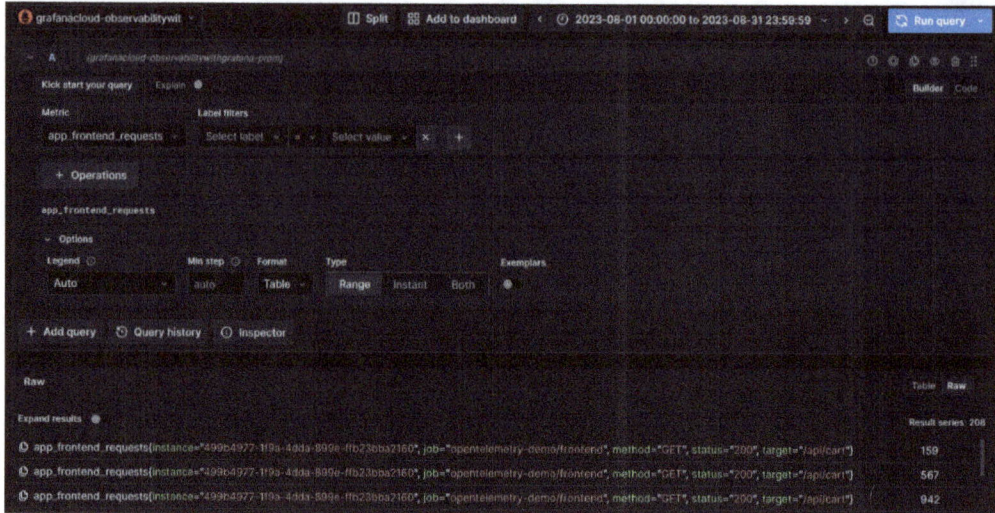

Figure 5.3 – Visualizing the samples and time series that make up a metric

Now that we understand the data structure, let's explore PromQL.

An overview of PromQL features

In this section, we will take you through the features that PromQL has. We will start with an explanation of the data types, and then we will look at how to select data, how to work on multiple datasets, and how to use functions. As PromQL is a query language, it's important to know how to manipulate data to produce alerts and dashboards.

Data types

PromQL offers three data types, which are important, as the functions and operators in PromQL will work differently depending on the data types presented:

- **Instant vectors** are a data type that stores a set of time series containing a single sample, all sharing the same timestamp – that is, it presents values at a specific instant in time:

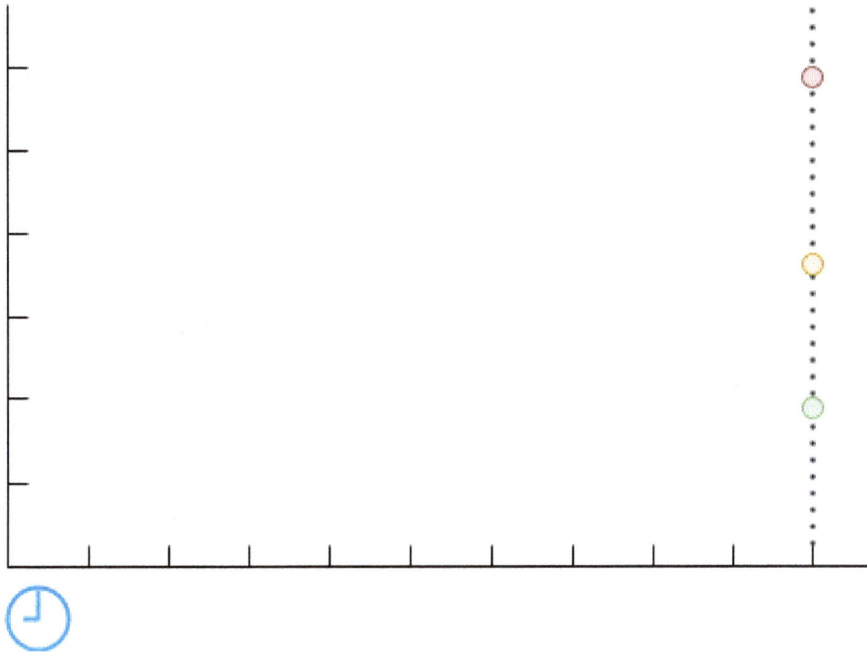

Figure 5.4 – An instant vector

- **Range vectors** store a set of time series, each containing a range of samples with different timestamps:

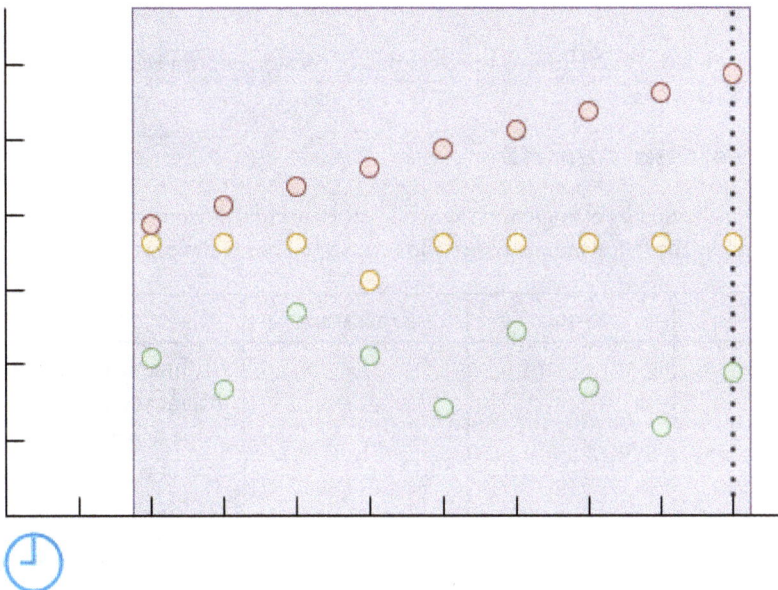

Figure 5.5 – Range vectors

- **Scalars** are simple numeric values, with no labels or timestamps involved.

Selecting data

PromQL offers several tools for you to select data to show in a dashboard or alert, or just to understand a system's state. Some of these are described in the following table:

Name	Syntax	Operators	Scope
Metric selector	`metric_name`		Selects a metric
Range selector	`[5m]`	ms, s, m, h, d, w, and y	Selects samples
Label selector	`{label="value", foo!="bar"}`	=, !=, =~, and !~	Selects and filters time series using labels
Offset modifier	`offset 5m`	ms, s, m, h, d, w, and y	Offsets the evaluation time from the current point in time by the specified amount
@ modifier	`@ 1686561123`	@	Sets the evaluation time to a specific time for instant or range vectors. This modifier uses epoch timestamps

Table 5.1 – The selection operators available in PromQL

In addition to the operators that allow us to select data, PromQL offers a selection of operators to compare multiple sets of data.

Operators between two datasets

Some data is easily provided by a single metric, while other useful information needs to be created from multiple metrics. The following operators allow you to combine datasets.

Name	Syntax	Operators	Scope
Arithmetic operators	`a + b`	+, -, *, /, %, and ^	Arithmetic operations on instant vectors and scalars; scope depends on the data type of a and b. It's important to note that vectors are matched on all labels.

Comparison operators	`a == b`	`==, !=, >, <, >=,` and `<=`	Filters instant vectors and scalars based on the comparison; scope depends on the data type of a and b.
Aggregation operators	`sum by (label) (a)`	`sum()`, `min()`, `max()`, `avg()`, `group()`, `stddev()`, `stdvar()`, `count()`, `count_ values()`, `bottomk()`, `topk()`, and `quantile()`	Aggregation operations on a single instant vector. These operators offer the `without` and `by` clauses to modify how results are grouped by label.
One-to-one vector matching	`a + on b`	`on()` and `ignoring()`	Modifies vector matching to specific labels (on) or ignoring a label (ignoring)
One-to-many/ many-to-one vector matching using group modifiers	`a + group_ left b`	`group_left()` and `group_ right()`	Modifies the vector matching in cases of many-to-one or one-to-many matching. Grouping can use a label list to include a label in the results.
Many-to-many vector matching using logical operators	`a and b`	`and, or,` and `unless`	Modifies vector matching in cases of many-to-many matching, based on logical operations between labels and the values of a and b

Table 5.2 – The comparison operators available in PromQL

Vector matching is an initially confusing topic; to clarify it, let's consider examples for the three cases of vector matching – *one-to-one*, *one-to-many/many-to-one*, and *many-to-many*.

By default, when combining vectors, all label names and values are matched. This means that for each element of the vector, the operator will try to find a single matching element from the second vector. Let's consider a simple example:

- **Vector A**:

    - `10{color=blue,smell=ocean}`

    - `31{color=red,smell=cinnamon}`

    - `27{color=green,smell=grass}`

- **Vector B**:

    - `19{color=blue,smell=ocean}`

    - `8{color=red,smell=cinnamon}`

    - `14{color=green,smell=jungle}`

- **A{} + B{}**:

    - `29{color=blue,smell=ocean}`

    - `39 {color=red,smell=cinnamon}`

- **A{} + on (color) B{}** *or* **A{} + ignoring (smell) B{}**:

    - `29{color=blue}`

    - `39{color=red}`

    - `41{color=green}`

When `color=blue` and `smell=ocean`, `A{} + B{}` gives `10 + 19 = 29`, and when `color=red` and `smell=cinnamon`, `A{} + B{}` gives `31 + 8 = 29`. The other elements do not match the two vectors so are ignored.

When we sum the vectors using `on (color)`, we will only match on the `color` label; so now, the two green elements match and are summed.

This example works when there is a *one-to-one* relationship of labels between vector **A** and vector **B**. However, sometimes there may be a *many-to-one* or *one-to-many* relationship – that is, vector A or vector B may have more than one element that matches the other vector. In these cases, Prometheus will give an error, and grouping syntax must be used. Let's look at another example to illustrate this:

- **Vector A**:

    - `7{color=blue,smell=ocean}`

- 5{color=red,smell=cinamon}

- 2{color=blue,smell=powder}

- **Vector B**:

  - 20{color=blue,smell=ocean}

  - 8{color=red,smell=cinamon}

  - 14{color=green,smell=jungle}

- **A{} + on (color) group_left B{}**:

  - 27{color=blue,smell=ocean}

  - 13{color=red,smell=cinamon}

  - 22{color=blue,smell=powder}

Now, we have two different elements in vector **A** with color=blue. The group_left command will use the labels from vector **A** but only match on color. This leads to the third element of the combined vector having a value of 22, when the item matching in vector **B** has a different smell. The group_right operator will behave in the opposite direction.

The final option is a *many-to-many* vector match. These matches use the logical operators and, unless, and or to combine parts of vectors **A** and **B**. Let's see some examples:

- **Vector A**:

  - 10{color=blue,smell=ocean}

  - 31{color=red,smell=cinamon}

  - 27{color=green,smell=grass}

- **Vector B**:

  - 19{color=blue,smell=ocean}

  - 8{color=red,smell=cinamon}

  - 14{color=green,smell=jungle}

- **A{} and B{}**:

  - 10{color=blue,smell=ocean}

  - 31{color=red,smell=cinamon}

- **A{} unless B{}**:

  - `27{color=green,smell=grass}`

- **A{} or B{}**:

  - `10{color=blue,smell=ocean}`

  - `31{color=red,smell=cinamon}`

  - `27{color=green,smell=grass}`

  - `14{color=green,smell=jungle}`

Unlike the previous examples, mathematical operators are not being used here, so the values of the elements are the values from vector **A**, but only the elements of **A** that match the logical condition in **B** are returned.

Now that we understand the operators, let's quickly introduce PromQL functions before we look at a practical example of writing PromQL. We will explore a practical example of their use in the *Writing PromQL* section.

Functions

PromQL offers about 60 different functions. The full list of functions can be found on the Prometheus website: `https://prometheus.io/docs/prometheus/latest/querying/functions`.

Now that we've looked at the functions available in PromQL, let's explore writing a PromQL query.

Writing PromQL

While technical descriptions of a language are useful for reference, this section will follow the process of building a query so that the language can be seen in context. Having your Grafana instance open in **Explorer** will help you follow along. In the following sections, we'll write practical examples using the selectors, operators, and modifiers we introduced in the previous section.

Metric selection

When we looked at metric labels, we saw how you can select metrics in PromQL with the `metric_name{}` syntax. This can be typed directly into a query using the **Code** button in the top-right corner of the query panel, or as we did earlier in the *Introducing PromQL* section, you can use **Builder**. In **Builder**, you will see the PromQL in the query panel below your selection, it should currently say `app_frontend_requests_total`. If it does not, use the **Metric** dropdown to select this metric. You should see results like in *Figure 5.3*. This method of selection returns an instant vector, as described in *Figure 5.4*.

The syntax is similar for returning a range vector, as described in *Figure 5.5*. We just need to add the range we are interested in – `metric_name[range]`. The range must include the time units, which can range from milliseconds (ms) to years (y). It's important to note that queries using range vectors need to be run with a query type of **Instant**. If a query type of **Range** or **Both** (the default) is selected, then you will receive an error. Here is an example of the error you will see:

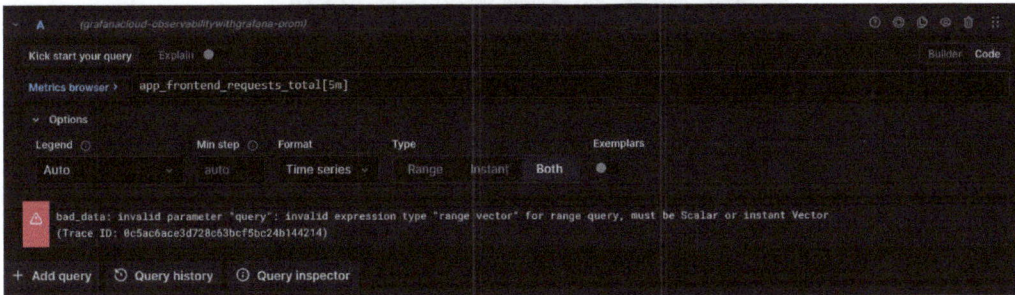

Figure 5.6 – An error when using a range vector in a range query

Time series selection and operators

As time series are made up of a unique set of labels, we can expand our query to only look at specific data – for example, only requests that target the cart API of the OpenTelemetry demo application. The following steps will filter our query to show only the requests that target the `/api/cart` endpoint:

1. Switch to the **Table** view at the top right of the **Results** panel:

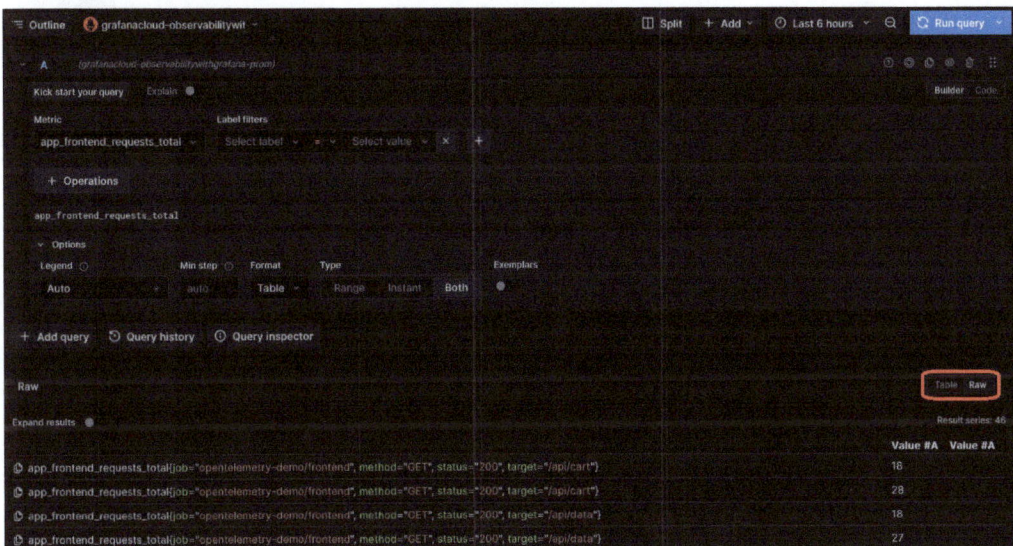

Figure 5.7 – Using the Table view in the PromQL results

2. Hover your mouse over a value for the target; you should see this icon:

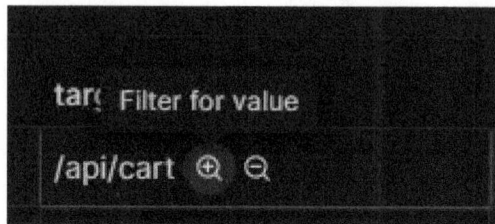

Figure 5.8 – Filter for value

3. Click on the plus icon, and you will see that we have a new label filter and our PromQL now says the following: `app_frontend_requests_total{target="/api/cart"}`.

4. Let's only show the `GET` method requests as well; you can do this by using the **filters** selector in the query panel, or the table in the **Results** panel. Like LogQL, we have different operators available to filter labels. The operators we can use are as follows:

    - `=`: Checks for an exact match of a string. For example, `target="/api/cart"` will match only when the `target` label is `/api/cart/`.

    - `!=`: Checks for anything other than an exact match. `target!="/api/cart"` will match everything except when the `target` label is `/api/cart/`.

    - `=~`: Checks for a regex match. For example, `target=~"/api/.*"` will match when the `target` label starts with `/api/`. This includes `/api/cart/`, `/api/horse/`, and `/api/cart/foo/bar/`.

    - `!~`: Checks for anything other than a regex match. `target!~"/api/.+"` will match when the `target` label is `/api/` or `/checkout/` but will not match `/api/cart/`, `/api/horse/`, and `/api/cart/foo/bar/`.

While we're looking at the table, you should also see a column titled **__name__**; this is a special label that can be used as an alternative during search, for instance, `metric_name{}` is equivalent to `{__name__="metric_name"}`.

We've now selected data and filtered it to the endpoint we're interested in, but a raw count of the requests that were made is difficult to interpret. Let's look at how to transform this count into something more useful, using a function.

Functions, aggregation, and operators

PromQL is a nested language, so to apply a function to a selected set of data, you simply enclose the data selection with the function. Our query so far looks like this:

```
app_frontend_requests_total{target="/api/cart",method="GET"}
```

This query returns the count of requests at each sample point. For most purposes, we are more interested in the rate of requests that hit that endpoint. This will allow us to answer questions such as what the peak rate is, or whether the rate is higher now or lower than at another point in time. The function to get this information is the `rate()` function. We can plug our current query into the function like this:

```
rate(app_frontend_requests_total{target="/api/cart",method="GET"}[$_
rate_interval])
```

The rate function takes an input of a range vector, so we have added the special `[$__rate_interval]` time variable. This is a Grafana feature that instructs Grafana to pick an appropriate interval, based on the scrape interval of the data source we have selected. This feature simplifies the technicalities of selecting the correct rate interval. A similar process is used for aggregation and other operators.

Now that we know how to get the rate of requests to the `/api/cart` endpoint, let's have a look at another example query.

HTTP success rate

A common **Service Level Indicator (SLI)** for a web application is the **success rate** of HTTP requests. In plain language, this is the number of successful HTTP requests/total HTTP requests. We will discuss the process of choosing good SLIs in *Chapter 9*.

A PromQL query like the following will produce the success rate SLI for the `app_frontend_requests_total` metric:

```
sum by (instance) (rate(
    app_frontend_requests_total{status=~"2[0-9]{2}"}[5m]))
/
sum by (instance) (rate(app_frontend_requests[5m]))
```

We can break this code down as follows:

- Using `app_frontend_requests_total{status=~"2[0-9]{2}"}[5m]`, we select samples of the `app_frontend_requests` metric that have the `status` label, with a value between `200` and `299`. This uses regex to select the label range, and it is a range vector over a five-minute range. For those of you familiar with regex, Grafana requires the escaping of backslashes.

- The `rate()` function calculates the per-second average rate of successful requests. This function returns an instant vector.

- The previous functions have left all data grouped into the initial time series from it. However, for this query, we are not interested in the method, target, or any other labels. Instead, we are interested in knowing whether a particular instance of the application is failing, as a failing instance could be masked by many good instances. To achieve this, we use the `sum by (instance) ()` aggregation.

- The last line of the query mirrors the first line but removes the label selector, so we get the *total* requests.

- Finally, we use the arithmetic operator (/) to divide the successful requests by the total requests. The output of this query gives us a number that will be close to 1 when most requests are successful; as we see failures, this will trend downward to 0 when every request fails.

Another common item to measure is the **duration** of requests made to the service. Durations are frequently represented as histogram data, and PromQL offers us many statistical tools we can use to understand our user's experience. Let's look at the following query:

```
histogram_quantile(
    0.95, sum(
        rate(
            http_server_duration_milliseconds_bucket{}[$__rate_
interval])
        ) by (le)
    )
```

The `http_server_duration_milliseconds_bucket` metric is a histogram, which is indicated by the naming convention of `_bucket`. The `histogram_quantile()` function takes this histogram data and gives us the 95th percentile duration. This is calculated using the `le` (less than or equal to) label in the histogram data. While it might be tempting to use averages for this kind of calculation, percentiles offer us a more nuanced understanding of the data. The 95th percentile means that 95% of samples have a duration less than or equal to the value returned.

Grafana offers several helpful functions to understand a query:

- Above the query component is a slider titled **Explain**. Toggling this on will present a step-by-step breakdown of what a query is doing.

- Also, above the query component is a button titled **Kick start your query**. Clicking this will give a number of starter queries.

- Below the **Options** section is the query **Inspector**. This will give detailed information about the query, such as its total request time and the data returned.

Here is a screenshot showing the location of these options:

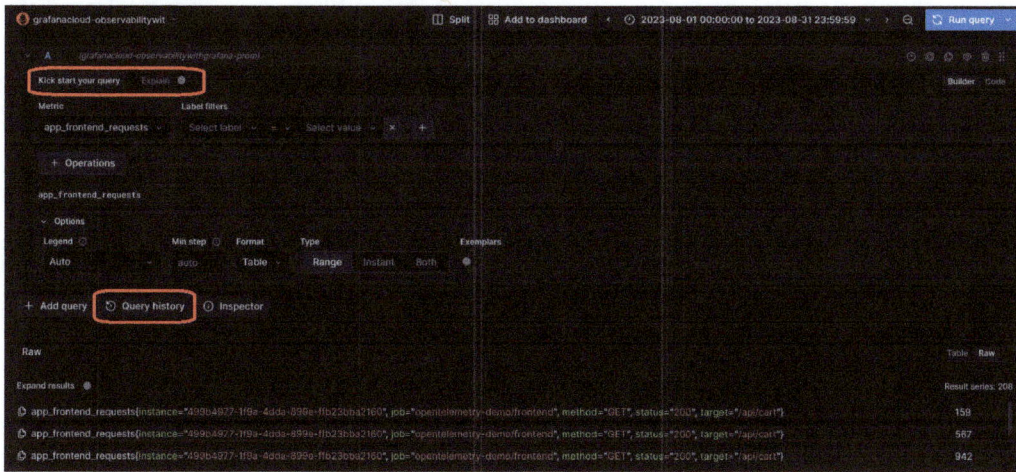

Figure 5.9 – Helpful functions for queries

Hopefully, you have a good grasp of the fundamentals of PromQL now and know what resources you have available to learn more. Whilst querying data is a major part of the day-to-day work in Grafana, it is good to have an understanding of how metrics data is collected.

The OpenTelemetry demo that has been set up also produces metrics from the single-node Kubernetes cluster, the kubelet instance on the node, and the underlying host. We encourage you to explore these metrics and see what you can find.

We've seen how to query the data stored in Prometheus-compatible systems. Now, let's see how to collect data from your services.

Exploring data collection and metric protocols

In *Chapter 2*, we introduced four common protocols in use to collect data from today's software – **StatsD** and **DogStatsD**, **OpenTelemetry Protocol** (**OTLP**), and **Prometheus**. We also introduced **Simple Network Management Protocol** (**SNMP**), which is used in the networking and compute spaces. In this section, we'll explore some of the features of these protocols.

There are two methods that metrics can be collected, push and pull. In a **push protocol**, the application or infrastructure must be configured with a destination to send metrics. In a **pull protocol**, the application or infrastructure is configured to expose metrics for another service to request. Both methods have advantages and disadvantages, it is also important to be aware of the potential security implications. In the following subsections, let's delve into each protocol.

StatsD and DogStatsD

We have grouped StatsD and DogStatsD together, as they are identical for the purposes of what we are discussing in this chapter.

StatsD is a *push protocol*, so each application producing metrics needs to be aware of the destination for these metrics. StatsD uses **User Datagram Protocol (UDP)** over port 8125 in its default settings. These are things to consider when using StatsD:

- StatsD uses UDP for transmission. This favors the speed of transmission over the guarantee of delivery.

- The protocol offers no support for authentication between the application and the receiving service. Depending on the environment, this could be a security concern.

It's worth noting that common practice, especially in Kubernetes, is to expose the StatsD receiver on localhost:8125, thus limiting exposure and offering a standard for applications to use.

StatsD has quite wide support in data collection agents, usually via contributed receivers. The OpenTelemetry collector, FluentBit, Vector, Beats, Telegraf, and the StatsD daemon all support the protocol. Prometheus offers an exporter that takes StatsD format metrics and exposes them as a Prometheus scrape endpoint; this is recommended as an intermediate step to a full Prometheus migration.

DogStatsD is less well supported than the StatsD format it is derived from; it provides an expanded set of metrics to StatsD. The data collection agents that natively support DogStatsD are **Vector** and Datadog's own agent. The OpenTelemetry collector currently has no support, but there are discussions in progress on adding this, and Datadog is an active participant in the OpenTelemetry project, so this is likely to change.

OTLP

OTLP is also a *push protocol*, so destination knowledge is necessary. Like StatsD, OTLP is often implemented using the standard receiving endpoint of localhost:4317 (**Google Remote Procedure Call (gRPC)**) or localhost:4318 (HTTP). OTLP supports both gRPC and HTTP and offers support for the authentication and acknowledgment between the client and server. OTLP also offers several quality-of-life items, such as **server-controlled throttling** and **GZIP compression**.

OpenTelemetry is in very active development, so this information is liable to change. As the project is a collaboration between several major vendors, agents from those vendors are increasingly supporting OTLP metrics. While other collection tools do not support OTLP input, the OpenTelemetry collector supports input from many sources. This means the OTEL collector is ideal for supporting a mixed estate. The vector collection agent also offers this versatility, and most things said about the OTEL collector can be applied to it as well.

Prometheus

Unlike StatsD and OTLP, Prometheus is a *pull protocol*. A client application needs to be configured to serve metrics on an endpoint, and then a Prometheus-compatible scraper is configured to collect those metrics at specific intervals. These metrics are commonly exposed on the `/metrics` endpoint, although some frameworks implement this differently (e.g., `/actuator/Prometheus` for Spring Boot).

It may seem that using a pull configuration increases the configuration steps required. However, using a pull method does reduce the information needed by the application of its running environment. For example, the application configuration would remain the same if 0 or 10 clients read its metrics. This pull pattern also matches very closely with the pattern of liveness and readiness endpoints for applications in Kubernetes.

To assist in the server configuration, Prometheus offers a wide range of service discovery options, across many different platforms, including Kubernetes, DNS, and Consul. These discovery options include matching a specific name and collecting data if a label is present, and this range of options allows for quite complex architectures where needed.

The Prometheus format has good collector support; Prometheus, the OTEL Collector, Grafana Agent, Vector, Beats, and Telegraf all support the collection of these metrics.

SNMP

SNMP is more complex than the other protocols discussed here, as it includes a lot of functionality for the management and monitoring of network-connected devices, such as switches and physical servers. The *monitoring* aspect of SNMP is a *pull protocol*, where a manager instance connects to agent software on devices and pulls data. There is additional functionality in **SNMP traps**, which allow a device to inform the manager about items as a data push. These traps are often of interest to track metrics from. It is worth noting that security can be a concern using SNMP, depending on how it is configured. SNMP offers a significant attack surface if configured incorrectly.

SNMP is very well supported, as the protocol has been active since 1988 and has good support from hardware vendors.

We've now covered querying data using PromQL, and how data is produced and collected, so let's now explore how Grafana stores metric data.

Understanding data storage architectures

Time Series Databases (**TSDBs**) are ideally suited to handle metric data, as metrics need to record data at specific points in time, and TSDBs are structured to make this data easy to record and query. There are several TSDBs available, but as this book is focused on Grafana, we will only discuss **Graphite**, **Prometheus**, and **Mimir** in this section. This is aimed at giving you an understanding of the structure of data as it is stored, as well as an overview of how Mimir allows organizations to scale their data beyond the capabilities of Graphite and Prometheus.

Graphite architecture

Graphite has several components; we will discuss the storage component **Whisper** here. The Whisper TSDB uses a flat file structure, where each unique time series is a fixed-size file. This size is determined by the configuration of resolution and retention configured in Whisper. Gathering this data for a search requires each of these files to be read, which quickly becomes expensive in disk I/O. As there are no inbuilt items that manage data redundancy, Graphite is also unable to guarantee that data written to it will be protected from loss or corruption.

However, the protocols introduced by Graphite to write data are still relevant although aging, so Grafana Cloud offers a Graphite ingest endpoint and query endpoint for teams that are already using this technology.

Graphite was an early example of metrics, introduced in 2008; the limitations of query speed and data integrity outlined previously led to the creation of Prometheus, which we will discuss next.

Prometheus architecture

Prometheus stores data in an immutable `block`, which covers a fixed time range (by default, two hours). Inside a block are several `chunks`, which are capped at 512 MB; these files contain the sampled value. Alongside these `chunks` are metadata files – `index` and `meta.json`. The `index` file contains a table that records the labels contained in the block and a reference to the position of all samples, with these labels in the associated chunks. Highly cardinal metric labels cause a huge increase in the size of the `index` file and degrade read performance. The `meta.json` file contains metadata such as the min and max timestamp contained in the `block` and stats on the samples, series, and chunks contained and the version used.

To process data as it's received, Prometheus also uses a `head block`, which is similar to the `block` used for storage, but it allows writes. This allows for the collection of a full two-hour block of data, ready for the index and metadata to be created when the block is finished. This process includes functionality to persist data on disk to prevent data loss. The `head block` consists of a **Write-Ahead Log** (**WAL**) that contains the raw data as it is received and a `meta.json` file that records what has been received. When the end of the two-hour time block is reached, a new `head block` is created, and the old `head block` is transformed into a standard `block`, with the creation of an `index` and `chunks`.

The following figure shows the structure of a fictional Prometheus TSDB, with the `blocks`, `chunks`, `index`, and metadata files and the WAL highlighted:

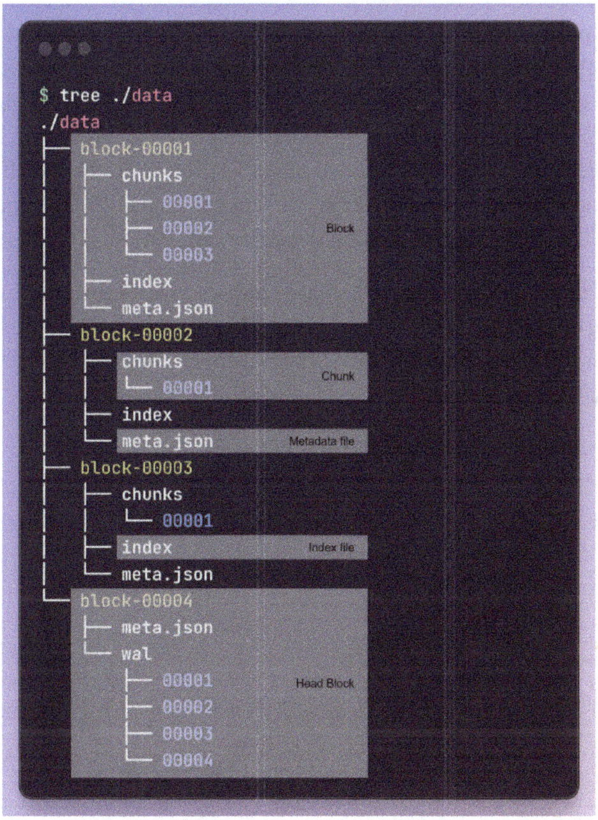

```
$ tree ./data
./data
├── block-00001
│   ├── chunks
│   │   ├── 00001
│   │   ├── 00002                Block
│   │   └── 00003
│   ├── index
│   └── meta.json
├── block-00002
│   ├── chunks
│   │   └── 00001                Chunk
│   ├── index
│   └── meta.json                Metadata file
├── block-00003
│   ├── chunks
│   │   └── 00001
│   ├── index                    Index file
│   └── meta.json
└── block-00004
    ├── meta.json
    └── wal
        ├── 00001                Head Block
        ├── 00002
        ├── 00003
        └── 00004
```

Figure 5.10 – The Prometheus TSDB

The implementation of the Prometheus TSDB in Prometheus itself is limited, as it uses local storage, which is not clustered or replicated natively. While it is possible to improve the aspects of this, there is a fundamental limitation of only a single node carrying out reads and writes. These limitations are perfectly acceptable in the correct circumstances. However, when scaling the TSDB to accept many active time series, changes are needed. Handling these situations is what Mimir was designed to do.

Mimir architecture

Mimir uses the same fundamental TSDB storage structures. However, unlike Prometheus, Mimir natively supports object stores for block files. The supported stores include Amazon S3, Google Cloud Storage, Microsoft Azure Storage, and OpenStack Swift.

By leveraging **object storage**, which is massively scalable, Mimir can handle the scaling problem experienced with Prometheus by adding new instances of the data-ingesting service. Mimir separates the incoming streams of data to a specific per-tenant TSDB, and each of these is assigned to an instance of the ingesting service. Like Prometheus, data is written to memory and the WAL by the ingester, and when the block is complete, it is written to object storage. To provide resilience Mimir will write each of these streams to multiple ingesters, and a compactor service will handle the process of merging the redundant blocks in object storage and removing duplicate samples.

Like the horizontal scalability of the write pathway, Mimir also scales the read pathway. It does this by splitting an incoming query into shorter time ranges. Then, it distributes these smaller units of the query to multiple querier instances. By doing this, Mimir again leverages the benefits of the underlying object storage for a quick return of data.

The following diagram shows the read and write pathways for Mimir:

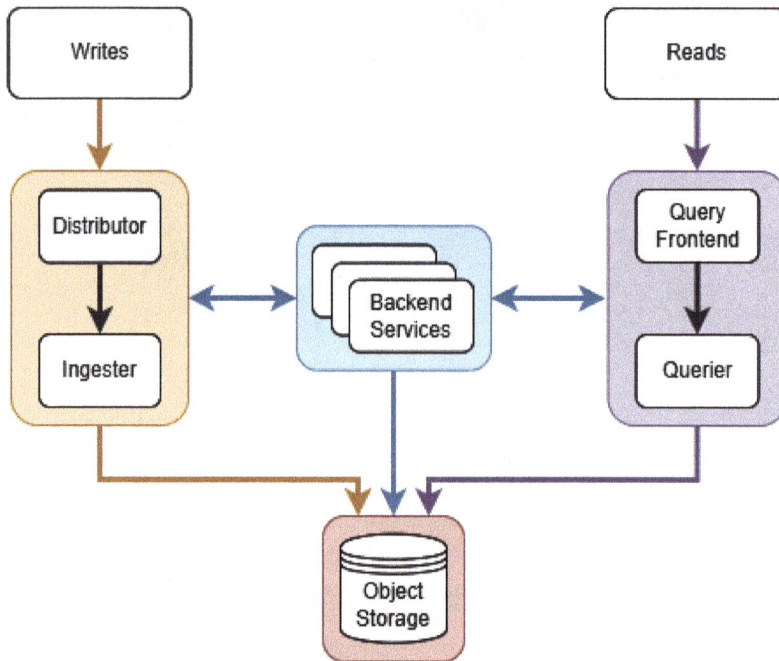

Figure 5.11 – Mimir architecture

Metrics show us aggregated data, such as the total count of requests. It is helpful when exploring an *odd* metric value to be able to look at an example. In applications that are instrumented with traces and metrics, exemplars allow us to record a sample trace in our metric data. Let's see this capability in action.

Using exemplars in Grafana

Exemplars are functions in Grafana that allow us to pivot from an aggregated view of the system, given by metrics, to a detailed view of a single request, given by traces. Exemplars need to be configured at the *collection layer* and then sent to the *storage layer*.

When they are available, you can view exemplars by doing the following:

1. Open **Options** under the query, and toggle the **Exemplars** slider:

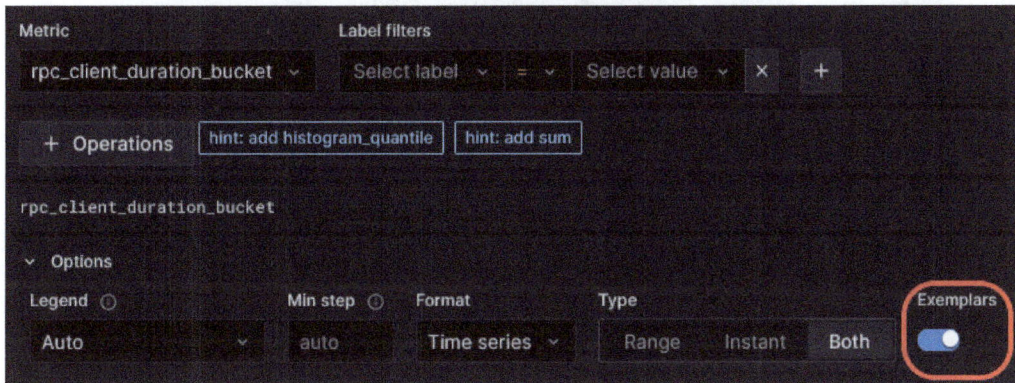

Figure 5.12 – The Exemplars toggle

2. Exemplars will appear as stars on the metrics chart:

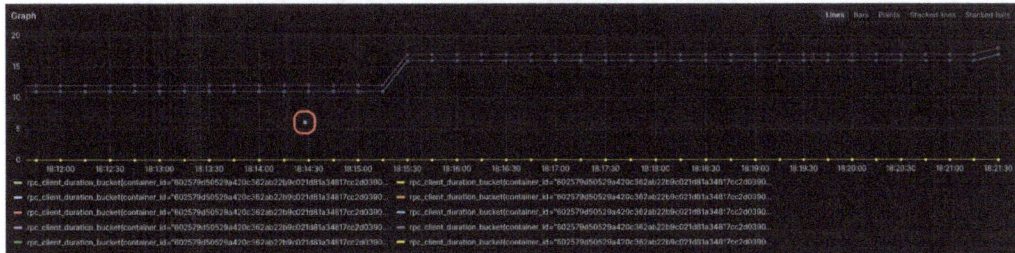

Figure 5.13 – An exemplar in metrics

Hovering over an individual exemplar will expand on the metrics data by showing information from the exemplar trace in the metrics view. We will explain these fields in more detail in *Chapter 6*, but some notable fields are the name and version of the process runtime and `span_id`, which would not usually be available in a purely metric view:

Exemplars	
Time	2023-08-19 18:14:27
Value	6.332472
__name__	rpc_client_duration_bucket
container_id	602579d50529a420c362ab22b9c021d81a34817cc2d0390cd3d98b85342878ca
host_arch	amd64
host_name	base-otel-demo-adservice-56f68d9b6c-s9hpf
instance	21679457-3164-4a46-ba0f-54a96fadb80a
job	opentelemetry-demo/adservice
k8s_namespace_name	default
k8s_node_name	k3d-owg-otel-demo2-server-0
k8s_pod_name	base-otel-demo-adservice-56f68d9b6c-s9hpf
le	10.0
net_peer_name	base-otel-demo-featureflagservice
net_peer_port	50053
net_transport	ip_tcp
os_description	Linux 5.15.90.1-microsoft-standard-WSL2
os_type	linux
process_command_line	/opt/java/openjdk/bin/java -javaagent:/usr/src/app/opentelemetry-javaagent.jar
process_executable_path	/opt/java/openjdk/bin/java
process_pid	1
process_runtime_description	Eclipse Adoptium OpenJDK 64-Bit Server VM 17.0.6+10
process_runtime_name	OpenJDK Runtime Environment
process_runtime_version	17.0.6+10
rpc_grpc_status_code	0
rpc_method	GetFlag
rpc_service	oteldemo.FeatureFlagService
rpc_system	grpc
service_instance_id	21679457-3164-4a46-ba0f-54a96fadb80a
service_name	adservice
service_namespace	opentelemetry-demo
span_id	dee234c6618da88b
telemetry_auto_version	1.24.0
telemetry_sdk_language	java

Figure 5.14 – Exemplar information

3. From an exemplar, you can also pivot from viewing metric data to looking at the trace in question by clicking on the **Query with Tempo (Tempo)** button:

Figure 5.15 – Opening an exemplar in Tempo

We'll discuss the details of tracing in more detail in *Chapter 6*, but this should give you a good introduction to using this kind of data in your metrics.

Summary

In this chapter, we explored metrics in detail. We saw all the operators available in PromQL and wrote two queries using the language. With that foundation of querying knowledge, we looked at the tools available to collect data and the various protocols with which applications can share data. We then looked at the architecture for Prometheu, and saw how Mimir takes the concepts of Prometheus and turns them into a highly scalable data processing tool, able to meet the needs of organizations of any size. Our final exploration was of Exemplars, giving us a concrete data example to add context to the aggregated data seen in metrics.

The next chapter will explore how traces work in Grafana Tempo, which will show you how powerful the use of exemplars and logging trace and span information can be to create a truly observable system for your organization's customers.

Tracing Technicalities with Grafana Tempo

Grafana Tempo is the third telemetry storage tool from Grafana that we'll discuss; it provides the capability to store and query trace data. This chapter will introduce the **Tempo query language** (**TraceQL**). TraceQL can be used to select and filter traces generated by your applications to gather insights from across traces; the language is very similar to LogQL and PromQL but tailored to trace data. In this chapter, we will explore the major tracing protocols and how they can be used to output traces from applications; this will help you make informed choices on which protocol to use in an application, or which protocols to support when collecting data. We'll then explore the architecture of Tempo to understand how it can fulfill the need for a scalable platform for tracing.

We will cover the following main topics in this chapter:

- Introducing Tempo and the TraceQL query language
- Exploring tracing protocols
- Understanding the Tempo architecture

Technical requirements

In this chapter, you will use the demo application and Grafana Cloud instance (set up in *Chapter 3*). You'll find the code for the chapter in the GitHub repository at https://github.com/PacktPublishing/Observability-with-Grafana/tree/main/chapter6. You'll find the *Code in Action* videos at https://packt.link/fJVXi.

Updating the OpenTelemetry Demo application

For this chapter, we have provided an updated OTEL-Collector.yaml with additional tracing configuration. This updated configuration is in the GitHub repository in the chapter6 directory. Full details on the update process are available from the GitHub repository in README.md.

To apply this updated configuration to the OpenTelemetry Collector, follow these steps:

1. Upgrade the Collector with Helm:

    ```
    $ helm upgrade --version '0.73.1' --values chapter6/OTEL-
    Collector.yaml --values OTEL-Creds.yaml owg open-telemetry/
    opentelemetry-collector
    NAME: owg-otel-collector
    LAST DEPLOYED: Sat Aug 19 12:42:36 2023
    NAMESPACE: default
    STATUS: deployed
    REVISION: 4
    ...
    ```

2. Validate that the upgrade was successful:

    ```
    $ kubectl get pods --selector=component=standalone-collector
    NAME    READY    STATUS    RESTARTS    AGE
    owg-otel-collector-594fddd656-tfstk    1/1    Terminating    1
    (70s ago)    2m8s
    owg-otel-collector-7b7fb876bd-
    vxgwg    1/1    Running    0    3s
    ```

    Your traces will now have more labels, and will also produce service graphs and span metrics.

Now that our local installation is updated, let's begin by exploring the third query language, TraceQL.

Introducing Tempo and the TraceQL query language

Tempo and TraceQL are the newest of the tools and query languages we will explore in depth in this book. Like LogQL, TraceQL was built using PromQL as an inspiration and offers developers and operators a familiar set of filtering, aggregation, and mathematical tools that aid in the observability flow between metrics, logs, and traces.

Let's have a quick look at how Tempo sees trace data:

- **Trace collection**: Introduced in *Chapter 2*, a trace (or distributed trace) is a collection of data that represents a request propagating through a system. Traces are often collected from multiple applications. Spans are sent by each application to some form of collection architecture and, ultimately, to Tempo for storage and querying.

- **Trace fields**: The following diagram introduces a simplified structure of a trace, similar to the simplified structure of logs, seen in *Chapter 4*, and traces, seen in *Chapter 5*:

```
trace_id: b5db4, span_id: 78a68, parent_id: null,
start_time: 2023-06-12T10:12:03.123123123Z,
end_time: 2023-06-12T10:12:12.123123123Z
```

Span 1

```
trace_id: b5db4, span_id: 571e5, parent_id: 78a68,
start_time: 2023-06-12T10:12:06.123123123Z,
end_time: 2023-06-12T10:12:12.123123123Z
```

Span 2

```
trace_id: b5db4, span_id: 885bb, parent_id: 78a68,
start_time: 2023-06-12T10:12:04.123123123Z,
end_time: 2023-06-12T10:12:10.123123123Z
```

Span 3

```
trace_id: b5db4, span_id: c91cd, parent_id: 885bb,
start_time: 2023-06-12T10:12:05.123123123Z,
end_time: 2023-06-12T10:12:13.123123123Z
```

Span 4

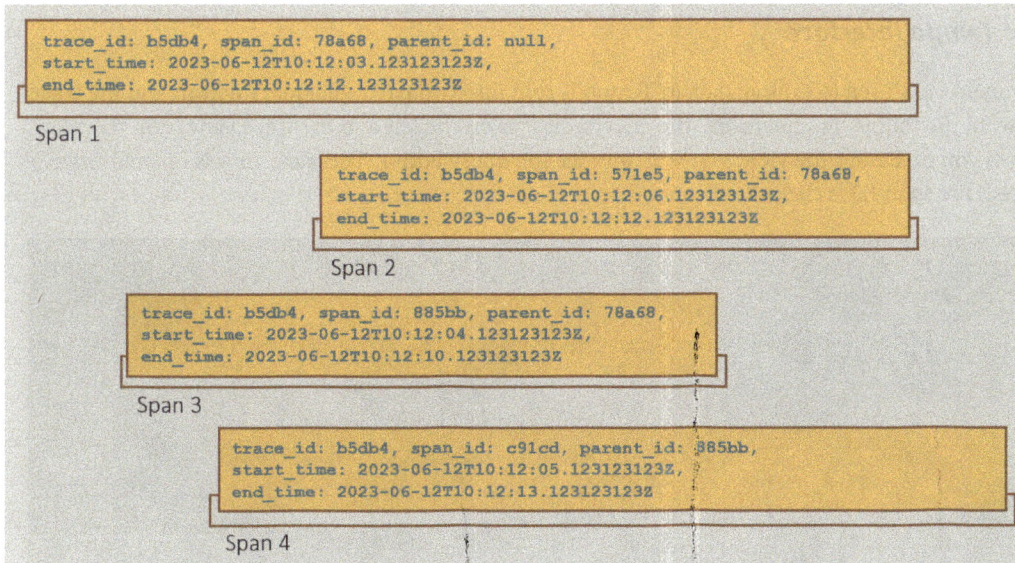

Figure 6.1 – A simplified view of a trace containing four spans

Back in *Chapter 2*, we introduced the common fields of a trace. In the preceding figure, we can
see that all four spans have the same `trace_id`, which is the unique identifier of the whole
trace. Each span has a unique identifier, the `span_id`. Each span also records where it came
from, using the `parent_id` field. Finally, the start and end times are recorded. This simplified
view does exclude several of the fields seen in the **OpenTelemetry Protocol** (**OTLP**), **Zipkin**,
and **Jaeger**, which are used to capture a lot of contextual information. We will discuss these
later in this chapter.

Now that we've seen the structure of trace data, let's now explore the Tempo interface and how we
can query data.

Exploring the Tempo features

In this section, we will introduce the major features of Tempo, the tracing platform available in
Grafana, and its query language, TraceQL. In *Chapters 4* and *5*, we introduced the LogQL and PromQL
languages, which focus on being able to select log or metric data and offer detailed functionality to
perform a powerful analysis of the selected data. Currently, PromQL only offers the ability to select
trace data. While there are powerful tools to select this data, there are no tools to perform an analysis.
Such functionality is an eventual aim for the product, but we wanted to highlight the current state of
Tempo at `v2.3.x`.

Let's begin by exploring the user interface for Tempo and how it represents trace data.

The Tempo interface

The main view used to explore data in Tempo is split into two parts, the **query editor**, and the **trace view**. In the following screenshot, the query editor is on the left and the trace view is on the right. When you first enter the view, you will only see the query editor. The trace view is opened when a trace ID or span ID is clicked on:

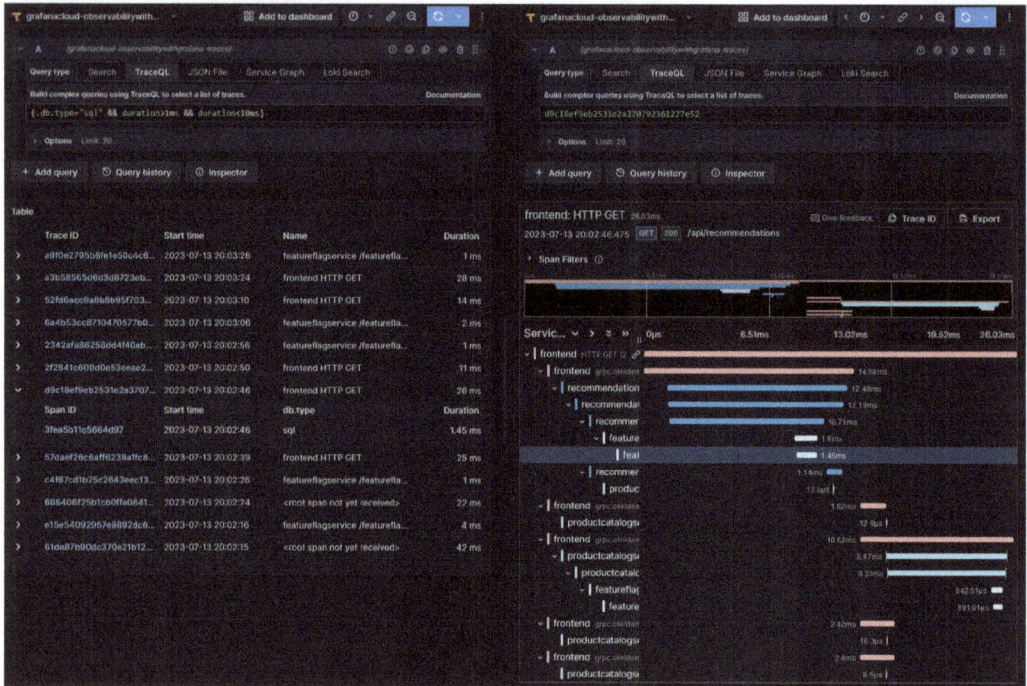

Figure 6.2 – The query editor (left) and the trace view (right)

The results panel is contextual. If we use TraceQL to run a search, it will return a list of traces and spans that match; this is shown on the left in the preceding screenshot. However, if we search for a specific trace ID, we will be shown the trace view on the right, where we can explore the spans in the trace.

While we are examining the query panel, let's look at two of the different search modes we can use, as shown here:

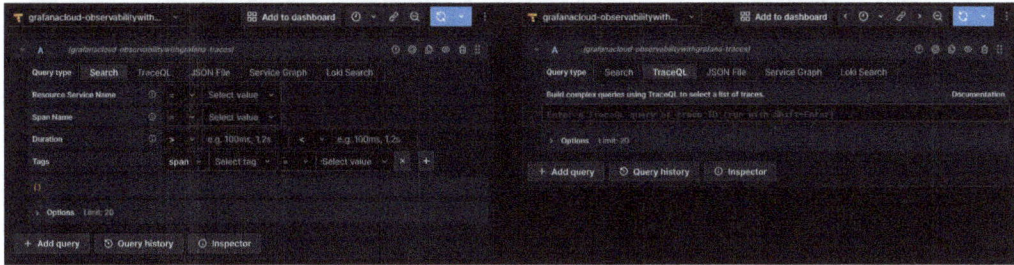

Figure 6.3 – Search modes

The two search modes shown in the preceding screenshot are as follows:

- **Basic Search mode**: Here, you are presented with drop-down menus to select the traces you are interested in. This is especially useful for people who are new to Tempo and want to get data quickly, but we will not explore this search mode in this book. Be aware that this mode is due to be deprecated in Grafana 10.3.

- **TraceQL mode**: This allows you to use TraceQL to search in a very granular way for the data you need. This is the default search mode.

As well as these, three search modes are available:

- **Loki Search mode**: This should be familiar to you from *Chapter 4*; it is available in Tempo, so you can pivot between logs containing trace and span IDs and a full trace view very quickly.

- **JSON File mode**: This allows for a trace saved in the JSON format to be imported and viewed directly. Combined with the export functionality, this allows for the simple preservation and sharing of interesting traces. Exploring the data in an exported JSON file is a good exercise for understanding the underlying data structures used in tracing.

- **Service Graph mode**: One of the most powerful features of collecting distributed traces is the ability to visualize the connections between those services. This tool gives anyone a clear graphical representation of how applications in a system communicate with each other. This functionality leverages metrics and traces together to represent a system's current state. The tool will also indicate erroring requests in red and successful requests in green.

The following screenshot shows the default view of a service graph:

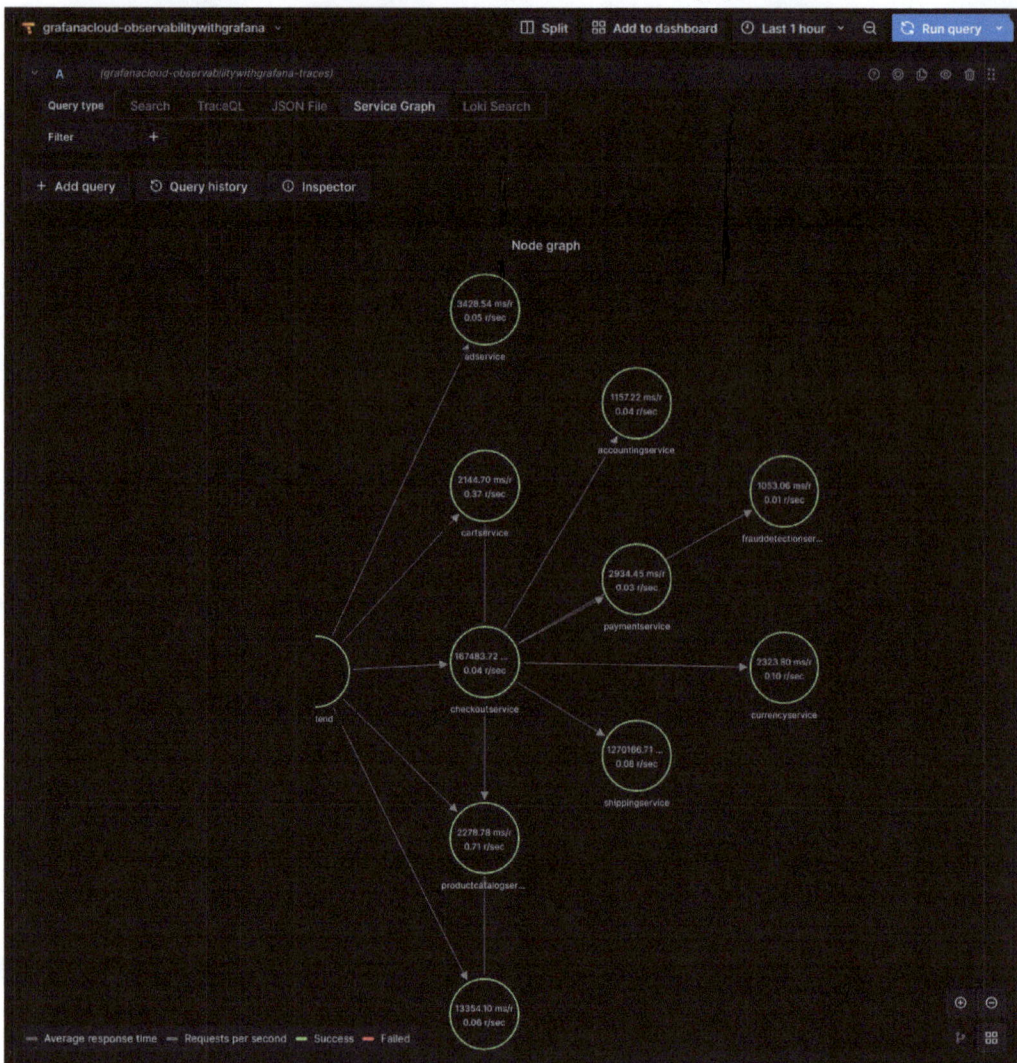

Figure 6.4 – Service graphs

As well as representing the connections between services, the preceding screenshot shows the request rates and average latency of the responses. Above the service graph, the **Requests, Errors, and Duration (RED)** metrics are shown. We will discuss these metrics in greater detail in *Chapter 9*.

At the time of writing, this aspect of OpenTelemetry and Tempo is under active development, and the authors are looking forward to the features that are coming.

Now that we have seen the interface for Tempo, let us understand how to use TraceQL to query trace data.

Exploring the Tempo Query language

Like Prometheus and Loki, Tempo offers a query language, **TraceQL**. Now that you are familiar with the interface of Tempo and the structure of traces, let's explore the features of TraceQL.

Field types

TraceQL uses two field types, **intrinsic fields** and **attribute fields**. Let's look at these in detail:

- **Intrinsic fields**: These are the fundamental information of spans and traces. These are used to show information in the trace view. The intrinsic fields are as follows:

  - `status`: The value could be `error`, `ok`, or unset (`null`)

  - `statusMessage`: Optional text to clarify the status

  - `duration`: The time between the start and end of the span

  - `name`: The operation or span name

  - `kind`: The value could be `server`, `client`, `producer`, `consumer`, `internal`, or `unspecified`, which is a fallback value

  - `traceDuration`: Number of milliseconds between the start and end of all spans in the trace

  - `rootName`: The name of the first span of the trace

  - `rootServiceName`: The name of the first service of the trace

- **Attribute fields**: These are the customizable fields that have been added to a span, either by the application or the collection tooling. Attribute fields are either span fields or resource fields, which is a distinction derived from the implementation of OpenTelemetry:

  - **Span attributes** are the fields added to the span by the submitting application; examples include `span.http.method` and `span.app.ads.ad_response_type`.

  - **Resource attributes**, conversely, represent the entity producing telemetry; examples include `resource.container.id` and `resource.k8s.node.name`.

  For efficient querying, it is best practice to always include `span.` and `resource.` in an attribute query. However, it is possible to use a leading `.` to query when you are unsure whether a field is a span or resource – for example, `.http.method` or `.k8s.node.name`.

When looking at an individual span, you can see the fields available under **Span Attributes** and **Resource Attributes**. This expanded view of a single trace shows the fields that are contained in the span:

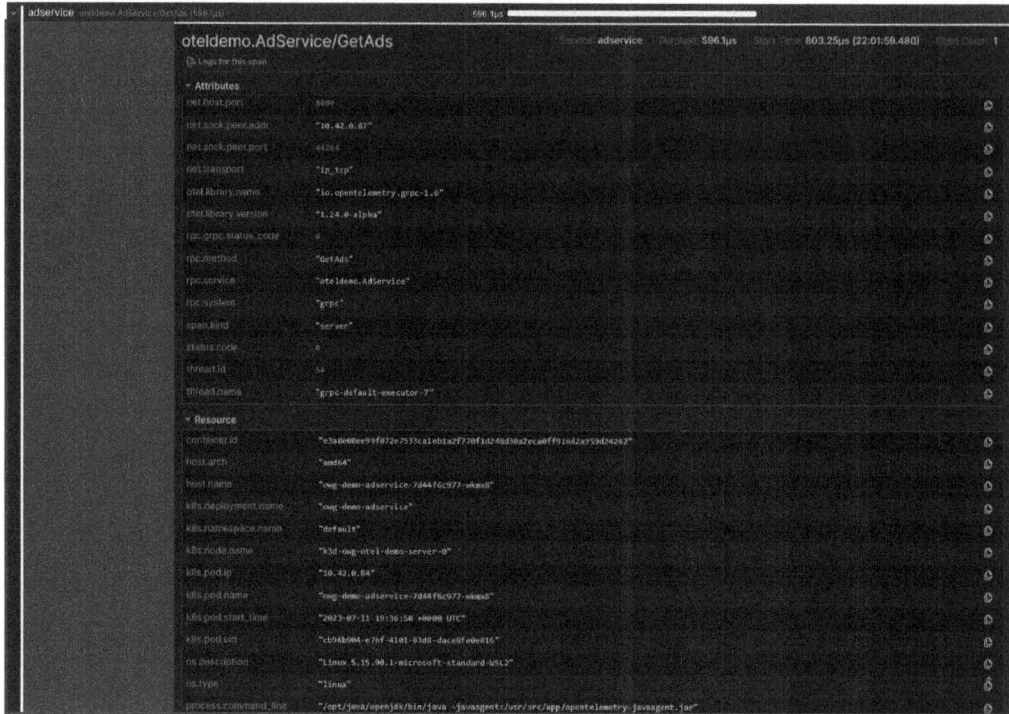

Figure 6.5 – Attributes for a span

Now that you have a good grasp of the fields available when searching traces in Tempo, let's have a look at how to search for traces and spans.

Selecting traces and spans

TraceQL offers tools to select data to show in a dashboard, or just to explore the current state of the system. These are described in the following table:

Name	Syntax	Operators	Scope
Field selector	`{field = " value"}`	=, ! =, >, >=, <, <=, =~, ! ~	Selects spans on the value of a field
Field expressions	`{field1="value1" && field2="value2"}`	&&, ||	Selects spans on the values of multiple fields

| Logical operators | `{field1="value1} && {field1="value2}` | `&&, \|\|` | Selects spans where a logical check between sets of spans is `true`. This can check multiple fields. |
| Structural operators | `{field1="value1"} > {field2="value2}` | `>, >>, ~` | Searches for spans in the second filter where they are related to the first filter.

These are explained in more detail after this table. |

Table 6.1 – The selection operators available in TraceQL

Structural operators offer the ability to carry out queries that take account of where conditions are met upstream (parent) or downstream (child) in a trace. Let's look at some examples:

- \> or the **child operator** refers to the direct child, such as the following:

```
{.service.name="frontend"} > {.service.
name="productcatalogservice"}
```

The preceding line would search for any span from the product catalog service, where the frontend service was the immediate parent.

- \>> or the **descendent operator** refers to any descendent, such as the following:

```
{.service.name="frontend"} >> {.service.name="cartservice"}
```

This would search for any span from the cart service where the frontend service was a parent in the trace but could have passed through another service first, such as the checkout service.

- ~ or the **sibling operator** refers to any spans that share the same parent, such as the following:

```
{.service.name="frontend"} ~ {.service.name="frontend"}
```

This would search for any span that visited the frontend multiple times. In the demo application, the frontend service would be the parent.

These operators allow us to select data. TraceQL also allows tools to carry out aggregation and mathematical functions on trace data.

Aggregators and arithmetic

Aggregators and mathematical functions allow for more complex queries. These can display information aggregated across all traces. Some of these are described in the following table:

Name	Syntax	Operators	Scope
Count aggregator	`| count() > 10`	`count()`	Refines the returned spans by the total count of spans in the span set
Numeric aggregators	`| avg(duration) > 20ms`	`avg()`, `max()`, `min()`, and `sum()`	Refines the returned spans by the field in the span set
Arithmetic operators	`{field1 < field2 * 10}`	`+, -, *, /,` and `^`	Performs arithmetic on numeric fields

Table 6.2 – Aggregation and mathematical operators in TraceQL

It is worth noting that TraceQL is in active development at the time of writing, so this list of operators is expected to grow.

Now that you have seen how to search trace data, let's discuss the important topic of moving seamlessly between data types to get a full picture.

Pivoting between data types

When correctly instrumented, an application will produce data that can be used to move between traces, logs, and metrics to truly understand what is happening.

Let's consider the following span, where an error was seen in `checkoutservice`. This could be a problematic error in a real shop, as it suggests a customer got to the checkout and was unable to complete their sale for some reason:

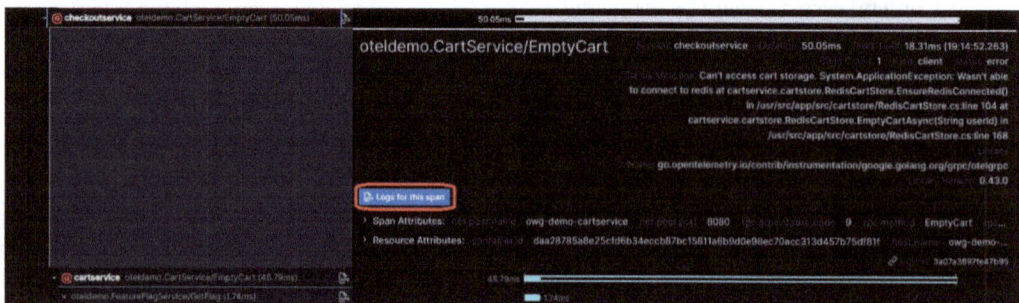

Figure 6.6 – Finding the logs for an error

The query interface for Tempo offers a helpful link, **Logs for this span**, which will open a Loki query. This functionality uses the `service_name` and `service_namespace` fields from the trace to query Loki. In a similar way, services can inject the trace context (`traceId` and `spanId`) into their log output where available. Loki can then be configured to provide contextual linking to Tempo, to see the trace view. Finally, as mentioned in *Chapter 5*, metrics can present exemplars, which allow users to see a sample trace from a metric graph.

We've explored the ways of seeing the data produced by applications. In the next section, we will understand the different protocols that are available to produce trace data for Grafana Tempo.

Exploring tracing protocols

In *Chapter 2*, we introduced the three main **tracing protocols**, OTLP, Zipkin, and Jaeger. In this section, we will explore some of the features of these protocols, how well-supported they are, and how to use them in the software services that you write. We will also discuss the different **headers** used by these protocols to propagate context to other services. A tracing protocol is made up of a set of headers that are added to the HTTP requests made by an instrumented application. These headers are what propagate the information of individual spans to downstream services. Once all of these spans are collected, they form a fully distributed trace.

What are the main tracing protocols?

First, let's look at the features and support of the main tracing protocols – OTLP, Zipkin, and Jaeger.

OTLP

OTLP tracing offers support for C++, .NET, Erlang, Go, Java, JavaScript, PHP, Python, Ruby, Rust, and Swift. There is good support for OTLP in popular development frameworks such as Spring, Django, ASP.NET, and Gin. With this wide support, it is best practice to search the documentation for your framework of choice on how to instrument an application; in most cases, instrumentation can be as simple as adding a few lines of dependencies.

Tracing is an inherently distributed process, and there have been several standards to propagate trace fields. This means that applications may need to use different HTTP or gRPC headers when handling traces, depending on other applications in their operating environment. OTLP provides native support for W3C TraceContext, B3, and Jaeger propagation headers, as well as support for W3C baggage headers, used to propagate other context information. The support of B3 and Jaeger headers means that applications instrumented with Zipkin and Jaeger libraries are natively supported. However, other trace headers such as AWS's X-Ray protocol are not maintained as part of the mainline distribution. If these protocols are used, it is recommended to use the relevant vendor's distribution of OpenTelemetry – for example, the *AWS Distro for OpenTelemetry* when X-Ray is used in a monitored environment (`https://aws.amazon.com/otel/`).

In the data collection space, OTLP trace data has good support from the OpenTelemetry Collector, Grafana Agent, FluentBit via a plugin, and Telegraf via a plugin.

Zipkin

Zipkin offers support for C#, Go, Java, JavaScript, Ruby, Scalar, and PHP via supported libraries, and C++, C, Clojure, Elixir, Lua, and Scala, via community-supported libraries. As with OTLP, there is also good support for Zipkin in popular development frameworks, so it is good practice to check the framework documentation when instrumenting applications.

Zipkin only natively supports the B3 propagation headers. However, as frameworks offer pluggable support for different trace protocols, support for alternative propagation headers is probably easy to implement in an application.

When it comes to data collection, Zipkin is supported by the OpenTelemetry Collector, Grafana Agent, and the native tools created by Zipkin.

Jaeger

We have included Jaeger for historical reasons here, but it is not recommended for adoption. Jaeger was originally developed by *Uber*. Before January 2022, Jaeger offered SDKs for Java, Python, Node.js, Go, C#, and C++. These SDKs supported the OpenTracing APIs. OpenTelemetry was formed by the OpenTracing and OpenCensus projects merging. Jaeger now recommends the use of the OpenTelemetry SDKs for instrumenting applications. For applications already using the Jaeger client libraries, migration guides have been provided by OpenTelemetry: `https://opentelemetry.io/docs/migration/opentracing/`.

Jaeger libraries supported the Jaeger, Zipkin, and W3C TraceContext headers, but they had no support for any other propagation formats.

There was not wide support in data collectors for Jaeger while it was actively supported; the intended way to use the protocol was to collect data in a Jaeger backend locally in an environment. The OpenTelemetry Collector and Grafana Agent do offer receivers for Jaeger traces and allow you to collect these traces as applications migrate to the OpenTelemetry protocol.

Now that you are familiar with the tracing protocols, let's look at the headers that are used to propagate information between services that use distributed tracing.

Context propagation

Distributed tracing is relatively new in web technologies, with the **World Wide Web Consortium (W3C)** making **Trace Context** a recommended standard in November 2021, while the **Baggage** format is still currently in a working draft state. Tracing records information in two distinct ways:

- Traces and spans are sent to a collection agent by each application

- Applications also share data using HTTP or gRPC headers, which are picked up by the receiving application

As tracing is a new technology, a couple of unofficial standard formats were used before the official W3C Trace Context headers were decided on. To provide some historical context on tracing, we'll explore the following formats:

- Jaeger/Uber headers

- Zipkin B3 headers

- W3C Trace Context headers

- W3C baggage headers

Jaeger/Uber headers

Jaeger libraries historically used the following header formats; we've included these for historical reference, as these should be considered deprecated in favor of W3C Trace Context.

The two HTTP headers used in Jaeger are `uber-trace-id` and `uberctx`, which look like this:

```
uber-trace-id: {trace-id}:{span-id}:{parent-span-id}:{flags}
uberctx-{baggage-key}: {baggage-value}
```

An example of the `uber-trace-id` header is as follows:

```
uber-trace-id: 269daf90c4589ce1:5c44cd976d8f8cd9:39e8e549de678267:0x01
```

Let's break this down the various fields in the `uber-trace-id` header:

- `trace-id`: This field is a 64-bit or 128-bit random number and is hex-encoded. In the example this is `269daf90c4589ce1`.

- `span-id` and `parent-span-id`: These are 64-bit random numbers and are hex-encoded. These are `5c44cd976d8f8cd9` and `39e8e549de678267`, respectively.

- `flags`: This field is used to convey additional information, such as whether the trace is being sampled. In this example, its value is `0x01`.

In the `uberctx` baggage header, the fields are as follows:

- `baggage-key`: This is a unique string that is used to name the header.

- `baggage-value`: This is a string that will be percent-encoded. Baggage as a concept will be explored further in the *W3C baggage* section.

Zipkin B3 headers

Zipkin libraries use B3 headers; unlike the Uber headers, Zipkin has historically separated each field into its own header, as shown in the following snippet:

```
X-B3-TraceId: {TraceId}
X-B3-ParentSpanId: {ParentSpanId}
X-B3-SpanId: {SpanId}
X-B3-Sampled: {bool}
X-B3-Flags: 1 OR header absent
b3: {TraceId}-{SpanId}-{SamplingState}-{ParentSpanId}
```

Let's break these headers down:

- `X-B3-TraceId`: Similar to the Jaeger format, `TraceId` is 64-bit or 128-bit hex-encoded. Here is an example of this header as it would be sent:

  ```
  X-B3-TraceId: 68720d6346a16000531430804ce28f9c
  ```

- `X-B3-ParentSpanId` and `X-B3-SpanId`: These are 64-bit and hex-encoded.

- `X-B3-Sampled`: This has a value of either `1` or `0`, although early implementations may use `true` or `false`.

- `X-B3-Flags`: This header is used to propagate debug decisions.

- `b3`: Zipkin predated the introduction of the W3C Trace Context standards. To aid in the transition to the newly agreed standard, Zipkin introduced the `b3` header. Later versions of Zipkin can propagate using both these headers and the W3C Trace Context headers for interoperability. The `b3` header exactly matches the `tracestate` header used in *W3C Trace Context* and represents the other headers combined into one mapping.

W3C Trace Context

W3C specifies a standard pair of headers, as shown here:

```
traceparent: {version}-{trace-id}-{parent-id}-{trace-flags}
tracestate: vendor specific trace information
```

Let's break these headers down into their constituents:

- The various fields in `traceparent` are as follows:

  - The `trace-id` is a hex-encoded 16-byte array (128-bit).

  - `parent-id` is a hex-encoded 8-byte array (64-bit); this field is equivalent to `span-id` or `SpanId` in Jaeger and B3, respectively, and represents the span ID used by the service that generated the header. It differs from the B3 `ParentSpanId`, as this can be used to represent a service further upstream that initiated a traced process.

- The `version` field is another hex-encoded 8-bit field; it represents the version of the standard being used. Currently, only version 00 exists.

- `trace-flags` is another hex-encoded 8-bit field. In version 00 of the W3C standard, the only available flag is one to denote whether sampling is occurring or not.

- `tracestate` is used to encode vendor-specific information. While `traceparent` is a fixed format and required by any vendor adopting the standard, `tracestate` is available for vendors to ensure that trace data is propagated while giving space for them to use and encode that data as desired. The only requirement regarding this header field is that the contents will be a comma-separated list of key-value pairs.

W3C baggage

Baggage is a related but different concept to a trace. Baggage headers contain contextual data that is passed between applications. These headers can share specific fields from one application and a downstream application. For example, we might have a top-level concept of `tenantId`, but when an application makes a request to a downstream application, that application may not need to know about `tenantId` for it to process the request. A baggage header allows us to propagate this `tenantId` field to the downstream application. The downstream application can then use this field in its observability instrumentation, while not polluting its data model with an unrelated field. This effectively separates *observability* concerns from *application* concerns. It's important to note that data contained in baggage headers can be exposed to anyone inspecting network traffic, so it should not be used to share sensitive information.

W3C baggage headers look like this:

```
baggage: key1=value1,key2=value2;property1;propertyKey=propertyValue,…
```

All fields must be percent-encoded; the full header must have 64 members or fewer, and it has a maximum size of 8,192 bytes. Using baggage gives systems a standardized way to propagate contextual information.

We've now discussed how trace data is produced by applications and how it is shared, both with a collection agent and other applications. Let's take some time to look at how data is processed and stored by Tempo.

Understanding the Tempo architecture

Like Loki and Mimir, Tempo leverages object stores such as Amazon S3, Google Cloud Storage, and Microsoft Azure Blob Storage. With the horizontal scalability of components in both the read and write pathways, Tempo has a fantastic ability to scale as data volumes increase.

The following diagram shows the architecture used by Tempo:

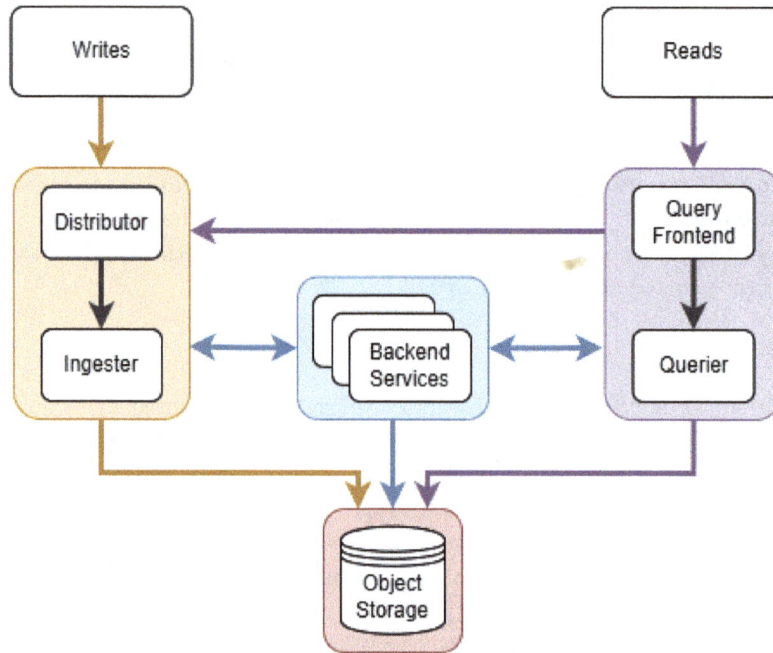

Figure 6.7 – The Tempo architecture

The *write* pathway for Tempo consists of the following:

- **Distributor**: The distributor is responsible for accepting spans and routing them to the correct instance of the ingester service, based on the trace ID of the span.

- **Ingester**: The ingester is responsible for grouping spans into traces, batching multiple traces into blocks, and writing bloom filters and indexes for querying. Once a block is complete, the ingester also flushes the data to the backend.

- **Metrics generator**: The metrics generator is an optional component; it receives spans from the distributor and uses them to produce service graphs and span metrics (such as the rate and the error duration). These are then written to a metrics backend.

The *read* pathway has these components:

- **Query frontend**: The frontend is responsible for receiving a query and splitting it into smaller shards, based on the blocks (created by the ingester) that will be read to return the requested data. These shards are then queued with queriers.

- **Querier**: This component is responsible for finding the requested data, either from the backend if the block has been flushed, or directly from the ingester if the block is still being collected.

The **compactor**, which is a standalone component, is responsible for optimizing the use of the backend storage.

Now that we're done exploring the system architecture of Tempo, you have seen all the major components of the tool, and how distributed tracing using Tempo can help provide great visibility of the components of the systems you run.

Summary

In this chapter, you learned how to use TraceQL to query trace data stored in Tempo, which will help you build queries for dashboards using this rich data source. You have explored the Tempo user interface, so you will be confident in moving around the interface. Combined with the skills learned in *Chapters 4* and *5* you will be confident in moving between log, metric, and trace data in Grafana to be able to observe the systems you work with.

We took a detailed look at the different protocols and libraries you can use when instrumenting an application, and we saw the levels of support those tools have across different programming languages. We also explored the HTTP headers that are used to propagate trace data between applications. This will help you choose the best way to instrument an application and how to work with applications that are already instrumented with tracing.

Finally, we looked at the Tempo architecture and how it can horizontally scale to support your organization with however many traces you need to sample. With this knowledge, you will understand how to operate a Tempo installation and monitor the various components.

In the next chapter, we will conclude *Part 2* of the book by showing you how to collect data from your infrastructure layers, whether that is a cloud provider such as AWS, Azure, or GCP, or a Kubernetes cluster.

7

Interrogating Infrastructure with Kubernetes, AWS, GCP, and Azure

This chapter will introduce the setup and configuration required to capture **telemetry** from various common cloud infrastructure providers. You will learn about the different options available for Kubernetes. Additionally, you will investigate the main plugins that allow Grafana to query data from cloud vendors such as **Amazon Web Services** (**AWS**), **Google Cloud Platform** (**GCP**), and Azure. You will look at solutions for handling large volumes of telemetry where direct connections are not scalable. The chapter will also cover options for filtering and selecting telemetry data before it gets to Grafana for **security** and **cost optimization**.

We will cover the following main topics in this chapter:

- Monitoring Kubernetes using Grafana
- Visualizing AWS telemetry with Grafana Cloud
- Monitoring GCP using Grafana
- Monitoring Azure using Grafana
- Best practices and approaches

Technical requirements

In this chapter, you will work with multiple cloud providers using a Grafana Cloud instance. You will need the following:

- A Grafana Cloud instance (set up in *Chapter 3*)
- Kubernetes and Helm (set up in *Chapter 3*)
- Accounts with the AWS, GCP, and Azure cloud providers with admin-level permissions

Monitoring Kubernetes using Grafana

Kubernetes has been designed to be monitored, and as such, it presents multiple options for anyone wanting to monitor it or the workloads running on it using Grafana. In this section, we will focus on monitoring Kubernetes, as we have already worked with Kubernetes workloads in previous chapters using the OpenTelemetry Demo application.

The OpenTelemetry Collector introduced in *Chapter 3* provides receivers, processors, and exporters to implement Kubernetes monitoring with data collection and enrichment. The following table identifies those components with a brief explanation for each of them:

OpenTelemetry Component	Description
Kubernetes Attributes Processor	The Kubernetes Attributes Processor appends Kubernetes metadata to telemetry, providing the necessary context for correlation.
Kubeletstats Receiver	The Kubeletstats Receiver obtains Pod metrics via a pull mechanism from the kubelet API. It collects node and workload metrics from each node it is installed on.
Filelog Receiver	The Filelog Receiver collects Kubernetes and workload logs that are written to `stdout` and `stderr`.
Kubernetes Cluster Receiver	The Kubernetes Cluster Receiver collects cluster-level metrics and entity events using the Kubernetes API.
Kubernetes Object Receiver	The Kubernetes Object Receiver collects objects for example events from the Kubernetes API.
Prometheus Receiver	The Prometheus Receiver scrapes metrics using Prometheus `scrape_config` settings.
Host Metrics Receiver	The Host Metrics Receiver scrapes metrics from Kubernetes nodes.

Table 7.1 – Kubernetes receivers

Let's now explore each component and how to implement them.

Kubernetes Attributes Processor

The OpenTelemetry **Kubernetes Attributes Processor** can automatically discover Pods, extract metadata from them, and add the extracted metadata to spans, metrics, and logs as additional resource attributes.

It provides necessary context to your telemetry, enabling the correlation of your application's metrics, events, logs, traces, and signals with your Kubernetes telemetry, such as Pod metrics and traces.

Data passing through the processor is by default associated to a Pod via the incoming request's IP address, but different rules can be configured.

The OpenTelemetry Collector Helm chart comes with several presets. For instance, the kubernetesAttributes preset, when enabled, will add the necessary RBAC roles to a ClusterRole and will add a k8sattributesprocessor to each enabled pipeline:

```
presets:
  kubernetesAttributes:
    enabled: true
```

Kubernetes comes with its own metadata to document its components. When using the kubernetesAttributes preset, the following attributes are added by default:

- k8s.namespace.name: The namespace the Pod is deployed to.

- k8s.pod.name: The name of the Pod.

- k8s.pod.uid: The unique ID for the Pod.

- k8s.pod.start_time: The timestamp for Pod creation, useful when understanding Pod restarts.

- k8s.deployment.name: The Kubernetes deployment name for the application.

- k8s.node.name: The name of the node the Pod is running on. As Kubernetes distributes the Pods over all of its nodes, it is important to understand whether any are having specific problems.

Additionally, the Kubernetes Attributes Processor creates custom resource attributes for your telemetry using Pod and namespace labels and annotations.

There are two methods applied to obtain and associate your data, that is, extract and pod_association. You can enable them in your Helm chart as detailed in the following code:

```
k8sattributes:
  auth_type: 'serviceAccount'
  extract:
  pod_association:
```

Let's look at these methods in greater detail:

- extract: This method provides the ability to use metadata, annotations, and labels as resource attributes for your telemetry. It has the following options:

  - metadata: Used to extract values from the Pod and namespace, such as k8s.namespace. name and k8s.pod.name

- annotations: Used to extract the value of a Pod or namespace annotation with a key and insert it as a resource attribute:

```
- tag_name: attribute-name
  key: annotation-name
  from: pod
```

- labels: Used to extract the value of a Pod or namespace label with a key and insert it as a resource attribute:

```
- tag_name: attribute-name
  key: label-name
  from: pod
```

Both annotations and labels can also be used with regex to extract part of the value for the new resource attribute.

- pod_association: This method associates data with the relevant Pod. You can configure multiple sources and the agent will try them in order, stopping when it finds a match. pod_association has the sources option, which is used to identify the resource attribute to use for the association, or it uses the IP attribute from the connection context:

```
pod_association:
  - sources:
      - from: resource_attribute
        name: k8s.pod.ip
  - sources:
      - from: resource_attribute
        name: k8s.pod.uid
  - sources:
      - from: connection
```

> **Permissions**
>
> If you are not using the kubernetesAttributes preset, you will have to provide the necessary permissions to allow access to the Kubernetes API. Usually, being able to access Pod, namespace, and ReplicaSet resources is adequate, but this will depend upon your cluster configuration.

Kubeletstats Receiver

The **Kubeletstats Receiver** connects to the kubelet API to collect metrics about the node and the workloads running on the node, which is why the preferred deployment mode is DaemonSet. Metrics are collected for Pods and nodes by default but can additionally be configured to collect metrics from containers and volumes.

The following code shows the configuration of the Kubeletstats Receiver:

```
receivers:
  kubeletstats:
    collection_interval: 60s
    auth_type: 'serviceAccount'
    endpoint: '${env:K8S_NODE_NAME}:10250'
    insecure_skip_verify: true
    metric_groups:
      - node
      - pod
      - container
```

Filelog Receiver

Although not a Kubernetes-specific receiver, the **Filelog Receiver** is the most popular log collection mechanism for Kubernetes. It tails and parses logs from files using operators chained together to process log data.

The OpenTelemetry Collector Helm chart has the `logsCollection` preset to add the necessary RBAC roles to the ClusterRole, and it will add a `filelogreceiver` instance to each enabled pipeline (we will explain `includeCollectorLogs` in *Chapter 10*):

```
presets:
  logsCollection:
    enabled: true
    includeCollectorLogs: false
```

If configuring this yourself, you will have to add the roles and `filelogreceiver` into your pipelines manually. A basic Filelog Receiver shows what to include and exclude, along with additional processing options:

```
filelog:
  include:
    - /var/log/pods/*/*/*.log
  exclude:
    - /var/log/pods/*/otel-collector/*.log
  start_at: beginning
  include_file_path: true
  include_file_name: false
```

Additionally, operators can be applied for log processing, filtering, and parsing.

The following is a list of Filelog Receiver parsers:

- `json_parser`: To parse JSON
- `regex_parser`: To perform regular expression parsing
- `csv_parser`: To parse comma-separated values
- `key_value_parser`: To process structured key-value pairs
- `uri_parser`: To process structured web paths
- `syslog_parser`: To process the standard syslog log format

Kubernetes Cluster Receiver

The **Kubernetes Cluster Receiver**, as its name suggests, collects metrics and events from the cluster using the Kubernetes API server. This receiver is used to obtain information regarding Pod phases, node conditions, and other cluster-level operations. The receiver must be deployed as a single instance; otherwise, the data would be duplicated.

An example cluster receiver configuration follows:

```
k8s_cluster:
  auth_type: serviceAccount
  node_conditions_to_report:
    - Ready
    - MemoryPressure
  allocatable_types_to_report:
    - cpu
    - memory
  metrics:
    k8s.container.cpu_limit:
      enabled: false
  resource_attributes:
    container.id:
      enabled: false
```

Kubernetes Object Receiver

The **Kubernetes Objects Receiver** can be used to collect any type of object from the Kubernetes API server. As with the Kubernetes Cluster Receiver, this must be deployed as a single instance to prevent duplicate data.

The receiver can be implemented to pull or watch objects by using `pull` or `watch`:

- When `pull` is implemented, the receiver periodically polls the Kubernetes API and lists all the objects in the cluster. Each object will be converted to its own log.

- When `watch` is configured, the receiver creates a stream with the Kubernetes API to receive updates as and when objects change; this is the most common use case.

Let's look at an example of Kubernetes Object Receiver configuration:

```
k8sobjects:
  auth_type: serviceAccount
  objects:
    - name: pods
      mode: pull
      label_selector: environment in (prod)
      field_selector: status.phase=Running
      interval: 15m
    - name: events
      mode: watch
      group: events.k8s.io
      namespaces: [default]
```

Prometheus Receiver

The **Prometheus Receiver** can be used to collect (scrape) metrics from Kubernetes and its workloads. The full range of Prometheus `scrape_config` options are supported by the receiver. An example of this implementation and `scrape_configs` can be seen in the *Chapter 5* demo project code. Here is an example Prometheus Receiver configuration:

```
prometheus:
  config:
    scrape_configs:
      - job_name: 'opentelemetry-collector'
        tls_config:
          insecure_skip_verify: true
        scrape_interval: 60s
        scrape_timeout: 5s
        kubernetes_sd_configs:
          - role: pod
```

The Prometheus Receiver is stateful, so the following points need to be taken into consideration when using it:

- The receiver cannot auto-scale the scraping process with multiple replicas

- Running multiple replicas with the same config will scrape targets multiple times

- To manually scale the scraping process, each replica will need to be configured with a different scraping configuration

Host Metrics Receiver

The **Host Metrics Receiver** collects metrics from a host using a variety of scrapers; the receiver will need access to the host filesystem volume to work correctly.

Table 7.2 shows the metrics available to scrape. The `OpenTelemetry Collector Helm chart` has the `hostMetrics` preset to add the necessary configurations:

```
mode: daemonset
presets:
  hostMetrics:
    enabled: true
```

By default, the preset will scrape every 10 seconds, which may generate too many metrics for your backend system. Be aware of this and consider overriding it to 60 seconds. The following table also shows the metrics that will be scraped by default using the preset:

Metric Scraper	Description	Included when using the hostMetrics preset
CPU	Scrapes CPU utilization metrics	Yes
Disk	Scrapes disk I/O metrics	Yes
Load	Scrapes CPU load metrics	Yes
Filesystem	Scrapes filesystem utilization metrics	Yes
Memory	Scrapes memory utilization metrics	Yes
Network	Scrapes network interface I/O metrics and TCP connection metrics	Yes
Paging	Scrapes paging and swap space utilization and I/O metrics	No
Processes	Scrapes process count metrics	No
Process	Scrapes per-process CPU, memory, and disk I/O metrics	No

Table 7.2 – Host Metrics Receiver scrapers

Let's now take a look at our first cloud provider, AWS, and the connectivity options available.

Visualizing AWS telemetry with Grafana Cloud

There are two main ways in which you can visualize your AWS telemetry with Grafana Cloud:

- **Amazon CloudWatch data source**: Amazon CloudWatch telemetry remains in AWS and Grafana is configured to remotely read the data at query time

- **AWS integration**: AWS CloudWatch telemetry data is either sent to or scraped and stored in Grafana Cloud (logs in Loki and metrics in Mimir).

Let's take a look at the differences between these two options to understand whether the integration option or the data source option best fits your use case.

Amazon CloudWatch data source

Grafana Cloud comes with support for **Amazon CloudWatch**, allowing you to query, trigger alerts, and visualize your data in Grafana dashboards. To read CloudWatch telemetry, you will need to configure the AWS **Identity and Access Management (IAM)** permissions and provide the necessary authentication details in the data source configuration screen. This does not store any telemetry data in Grafana; it only retrieves it at query time.

Let's now look at the different configuration steps.

Configuring the data source

Data sources can be accessed from the menu under the **Connections** item. To create a new connection, click the **Add new data source** button and search for CloudWatch. For existing ones, search for CloudWatch in the **Data sources** search box. You will see a screen similar to the following. Click **CloudWatch** to open the **Settings** page:

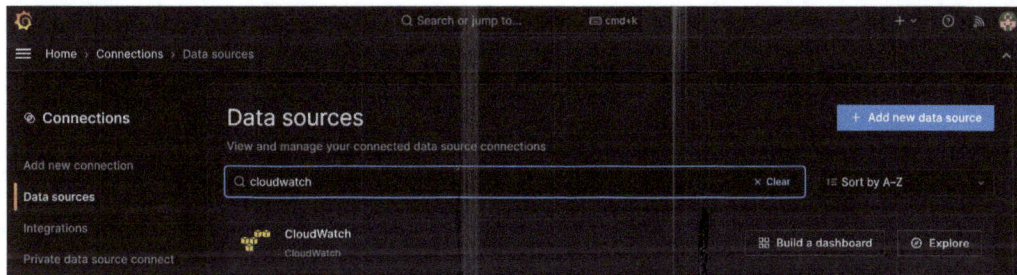

Figure 7.1 – Grafana Connections Data sources screen

The **Settings** screen requires the AWS configuration details needed to establish the connection, as shown in the following screenshot. Here, you can also configure namespace details for custom metrics, log query timeouts, and X-Ray trace links:

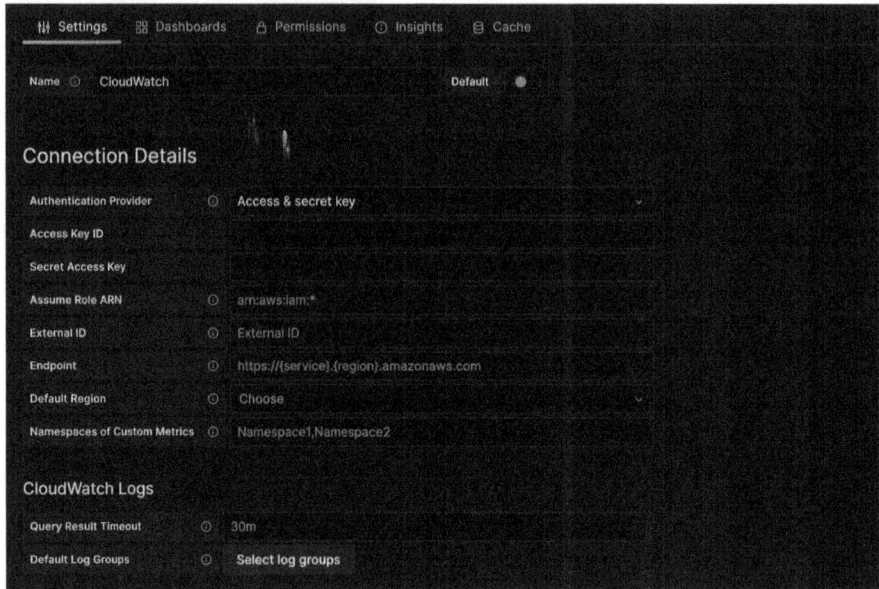

Figure 7.2 – Amazon CloudWatch data source settings

Using the Amazon CloudWatch query editor

The CloudWatch data source comes with its own specialized query editor that can query data from both CloudWatch metrics and logs.

From the data explorer, you can select **CloudWatch Metrics** or **CloudWatch Logs** as the source data, as shown in the following screenshot:

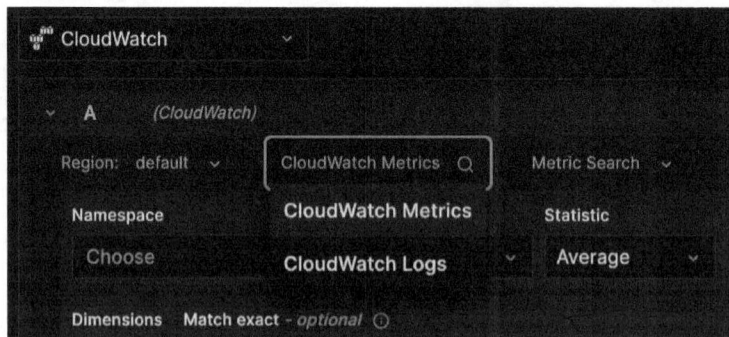

Figure 7.3 – Amazon CloudWatch query editor

With the metrics editor in **Builder** mode, you can create a valid metric search query by specifying the namespace, metric name, and at least one statistic.

The logs editor provides a **Log group** selector, allowing you to specify the target log groups and then use AWS CloudWatch Logs Query Language `https://docs.aws.amazon.com/AmazonCloudWatch/latest/logs/CWL_QuerySyntax.html` in the query editor.

Using Amazon CloudWatch dashboards

On the data source **Settings** screen, there is a **Dashboards** tab with a set of pre-configured dashboards to get you started.

The following figure shows the list of available dashboards and import details (if a dashboard has already been imported, you will see the options to delete or reimport):

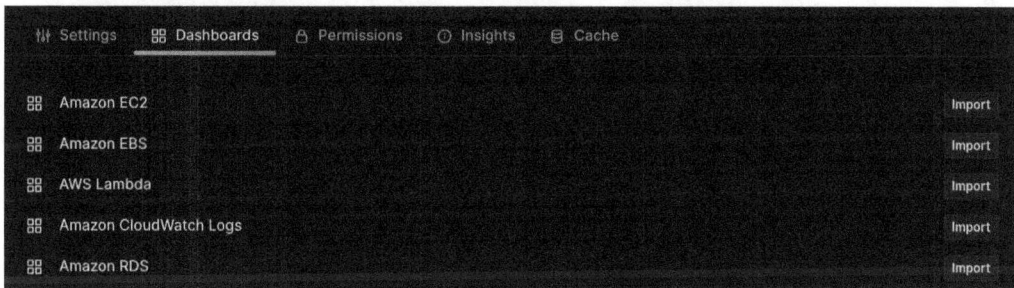

Figure 7.4 – Amazon CloudWatch pre-configured dashboards

Let's now take a look at the AWS integration option.

Exploring AWS integration

The AWS integration option can be added to your account. It will then be available as a connection. When added and configured, you will be able to ingest metric and log data directly into Grafana, which provides a query time benefit as the data is contained within your Grafana Cloud stack. The metrics and logs can then be queried using LogQL or PromQL; see *Chapter 4* and *Chapter 5* for refreshers.

Let's now look at the different configuration steps.

Configuring the integration

The AWS connection can be accessed from the menu under the **Connections** item. Search for `aws` from the **Add new connection** screen; you will see a screen similar to the following. You will see this is an **Infrastructure** connection and is labeled as **Guide**. This means there will be comprehensive instructions to help you connect the account and walk you through the process:

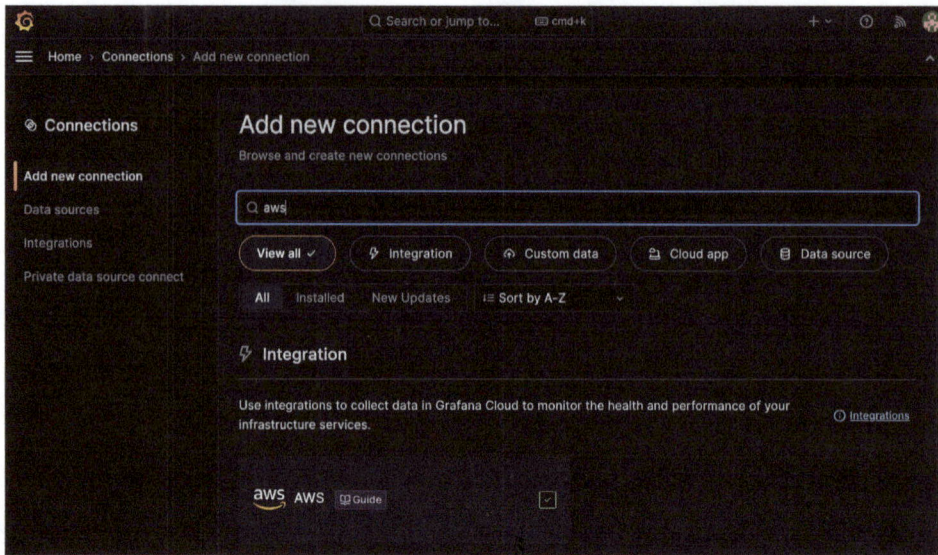

Figure 7.5 – Grafana Add new connection screen

Selecting the AWS integration option presents you with several options for integration – **CloudWatch metrics**, **Logs with Lambda**, and **Logs with Firehose** – as shown in the following screenshot:

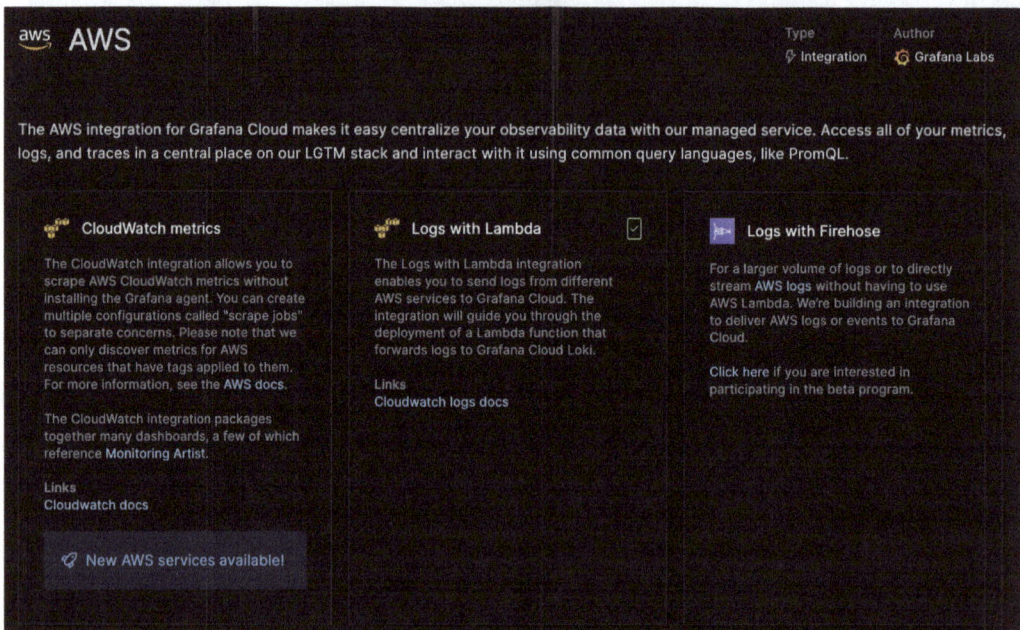

Figure 7.6 – AWS integration screen

Next, we will discuss CloudWatch metrics and Logs with Lambda.

CloudWatch metrics

The CloudWatch integration allows you to scrape Amazon CloudWatch metrics without installing any collector or agent infrastructure. Multiple scrape jobs can be created to separate concerns, but metrics can only be discovered for AWS resources with tags.

As mentioned earlier, this integration is guided and presents you with all the necessary details to get started by using infrastructure as code or by manually connecting and configuring scrape jobs. The following screenshot shows the CloudWatch metrics **Configuration Details** screen:

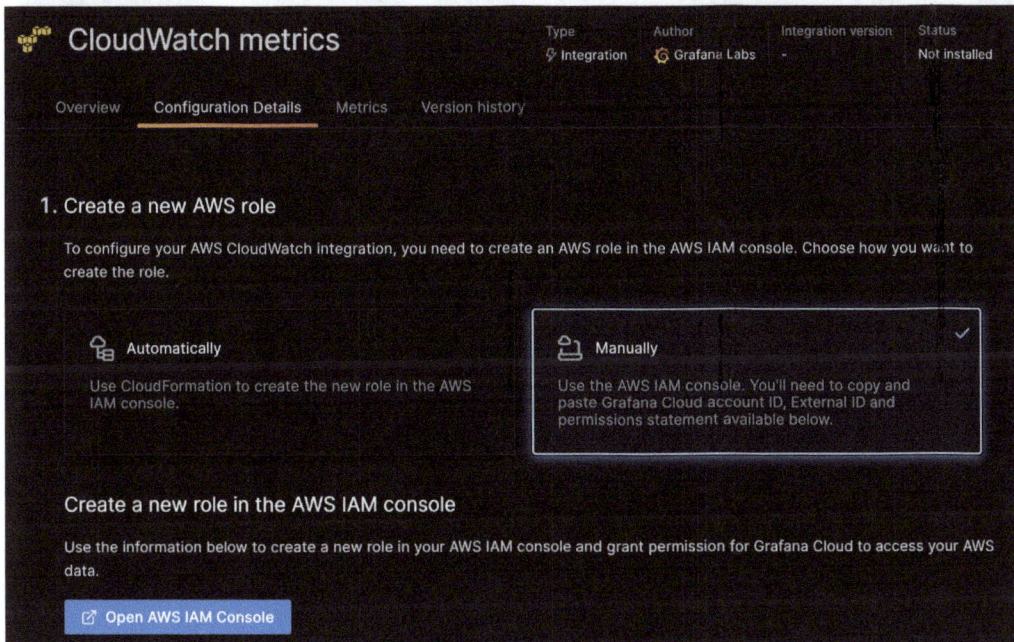

Figure 7.7 – CloudWatch Configuration Details screen

Additionaly there are some pre-built dashboards that are ready to use. The following figure shows the list of pre-built dashboards that come with the integration option at the time of writing:

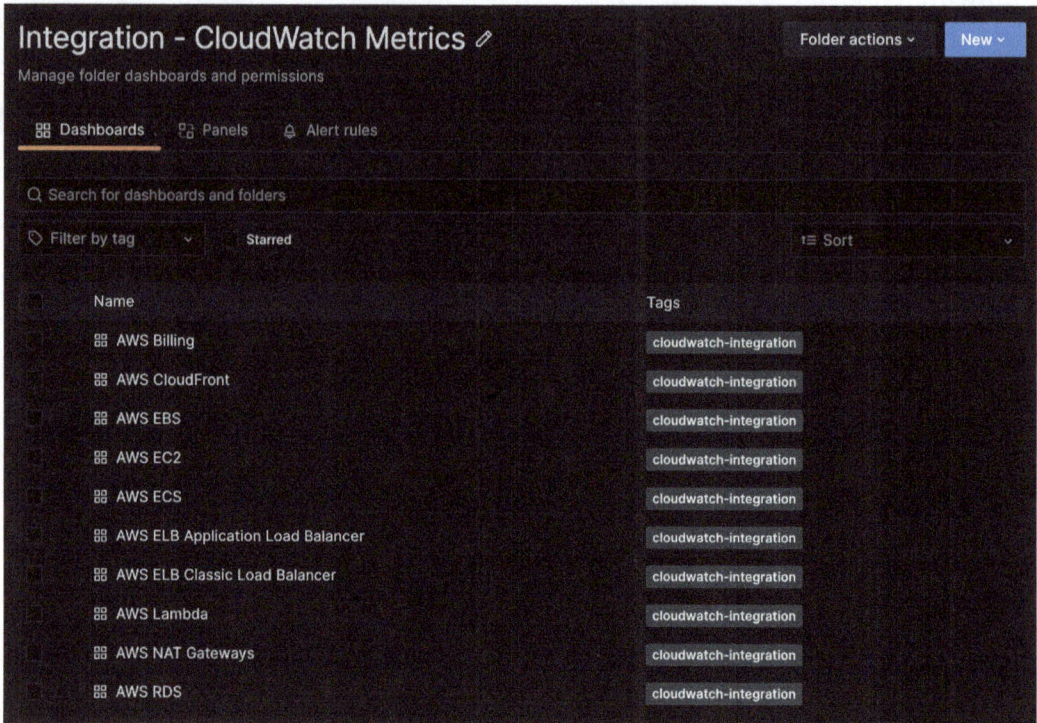

Figure 7.8 – Sample CloudWatch Metrics dashboard

Logs with Lambda

The Logs with Lambda integration enables you to send CloudWatch logs to Grafana Cloud. The integration will guide you through the deployment of an AWS Lambda function that forwards CloudWatch logs to Grafana Cloud Loki, where they can be queried using LogQL. *Chapter 4* explains Loki and LogQL in detail.

The following screenshot shows the **Logs with Lambda** configuration screen where you can select your deployment approach:

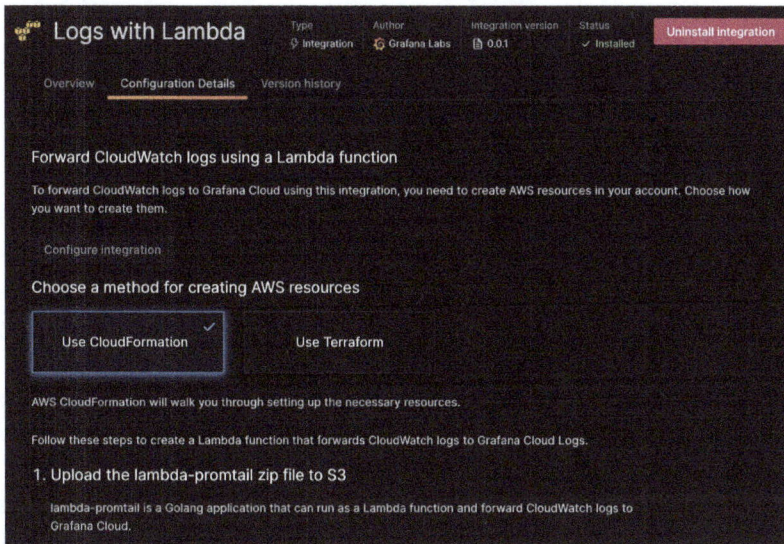

Figure 7.9 – Logs with Lambda configuration details

The following screenshot shows the configuration steps as the onscreen guide walks you through the connection and configuration of the logs integration:

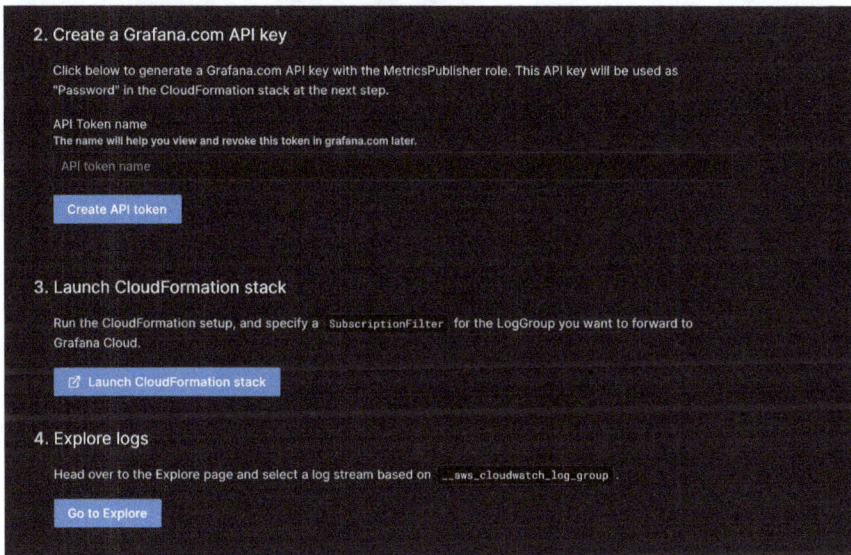

Figure 7.10 – Logs with Lambda CloudFormation configuration

Let's now look at our second cloud provider – GCP.

Monitoring GCP using Grafana

Grafana Cloud comes with support for **Google Cloud Monitoring**, allowing you to query, trigger alerts, and visualize your data in Grafana dashboards. It does not store any telemetry data in Grafana; it only retrieves it at query time.

Let's now look at the steps for configuring the data source.

Configuring the data source

Data sources can be accessed from the menu under the **Connections** item. Search for Google Cloud Monitoring in the **Data sources** search box; you will see a screen similar to the one shown in *Figure 7.11*. Click on **Google Cloud Monitoring** to open the settings page. The settings screen prompts for the Google configuration needed to establish and test the connection. The **Connections** search results screen is shown here:

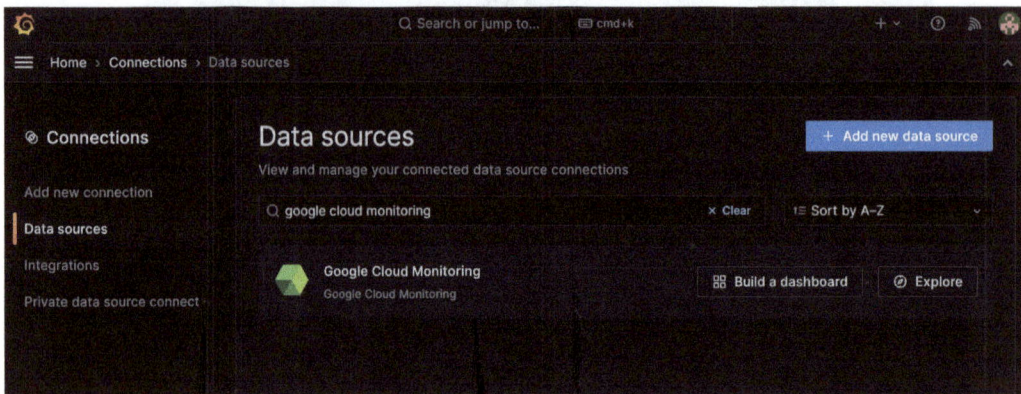

Figure 7.11 – Connections search results screen

The configuration settings for **Google Cloud Monitoring** shown in the following screenshot walk you through the configuration, helping you to choose an authentication method of either **JSON Web Token (JWT)** or **GCE Default Service Account**:

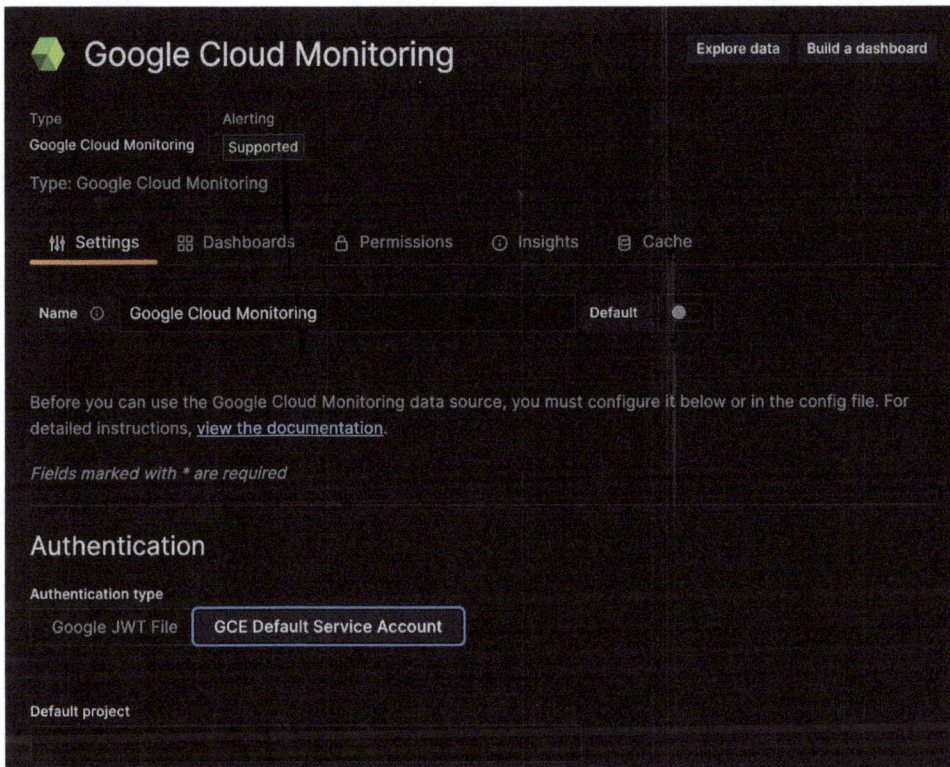

Figure 7.12 – Google Cloud Monitoring configuration settings

Depending upon the size of your GCP deployment, you may have to consider, as part of your design, any limits imposed on the token or service account.

Google Cloud Monitoring query editor

The Google Cloud Monitoring data source comes with its own specialized query editor that can help you build queries for metrics and GCP **Service Level Objectives** (**SLOs**), both of which return time-series data (you will learn more about visualizing time-series data in *Chapter 8*). Metrics can be queried using the **Builder** interface or using GCP's **Monitoring Query Language** (**MQL**). The SLO query builder helps you visualize SLO data in a time-series format. To understand the basic concepts of GCP service monitoring, refer to the GCP documentation at `https://cloud.google.com/stackdriver/docs/solutions/slo-monitoring`.

The **Google Cloud Monitoring** query editor in the following screenshot shows the three available choices:

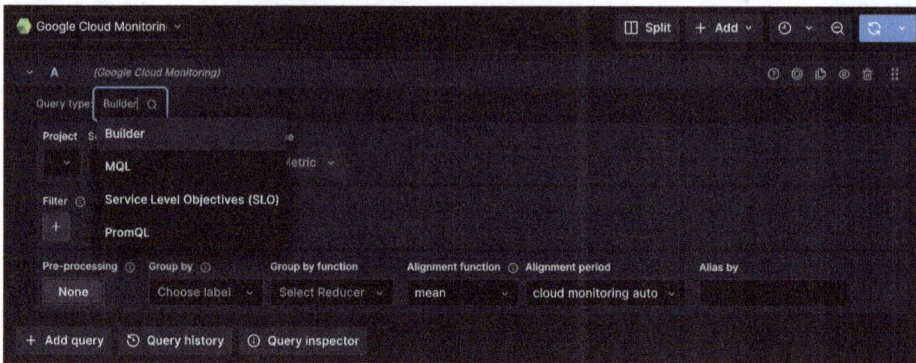

Figure 7.13 – Google Cloud Monitoring query editor selection

Let's look at the different Explorer query types:

- **Metrics queries**: The metrics query editor builder helps you select metrics, group and aggregate them by labels and time, and specify time filters for the time-series data you want to query:

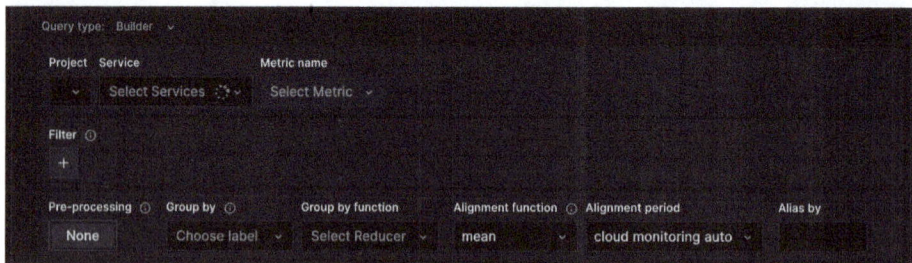

Figure 7.14 – Google Cloud Monitoring query editor metrics builder

The following screenshot shows the query editor for MQL, which provides an interface to create and execute your MQL query:

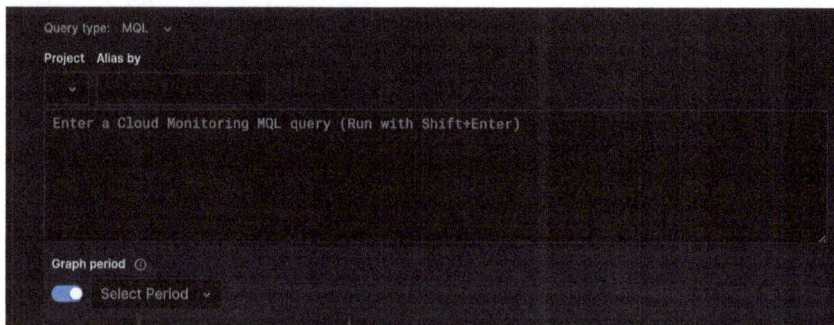

Figure 7.15 – Google Cloud Monitoring query editor metrics MQL interface

Full documentation for the MQL language specification can be found on the Google Cloud website at `https://cloud.google.com/monitoring/mql`.

- **SLO queries**: The SLO query builder helps you visualize SLO data in time-series format. Documentation to explain the basic concepts of service monitoring can be found on the Google Cloud website at `https://cloud.google.com/stackdriver/docs/solutions/slo-monitoring`.

The Google Cloud Monitoring SLO query editor is shown in the following screenshot:

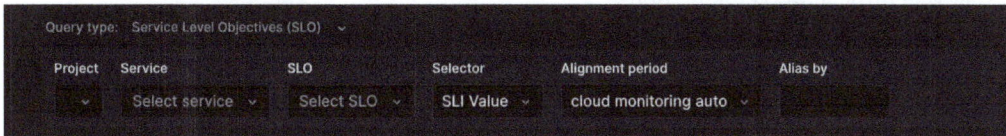

Figure 7.16 – Google Cloud Monitoring query editor metrics SLO builder

Google Cloud Monitoring dashboards

From the **Data source | Settings** screen, the **Dashboards** tab lists a set of pre-configured dashboards to get you started. The following screenshot shows the list of available dashboards at the time of writing and import details (if a dashboard has already been imported, there are options to delete or reimport). You can see from the list the various GCP components that are covered, including firewalls, data processing, SQL, and so on:

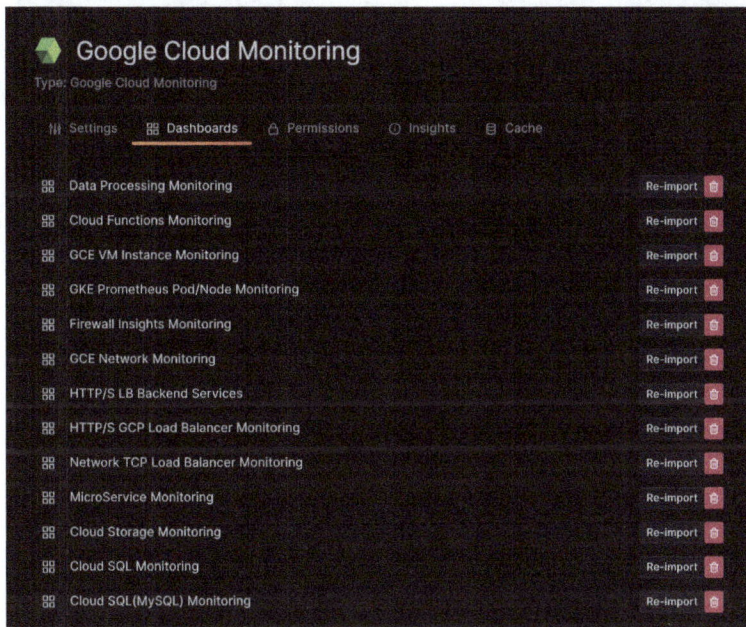

Figure 7.17 – Google Cloud Monitoring pre-built dashboards

Let's now look at our third cloud provider, Azure.

Monitoring Azure using Grafana

Grafana Cloud comes with support for Azure, allowing you to query, trigger alerts, and visualize your data in Grafana dashboards. This is called the **Azure Monitor** data source. As with the other cloud data sources, it does not store any telemetry data in Grafana; it only retrieves it at query time.

Let's now step through the configuration.

Configuring the data source

Data sources can be accessed from the menu under the **Connections** item. Search for Azure Monitor in the **Data sources** search box; you will see a screen similar to the following:

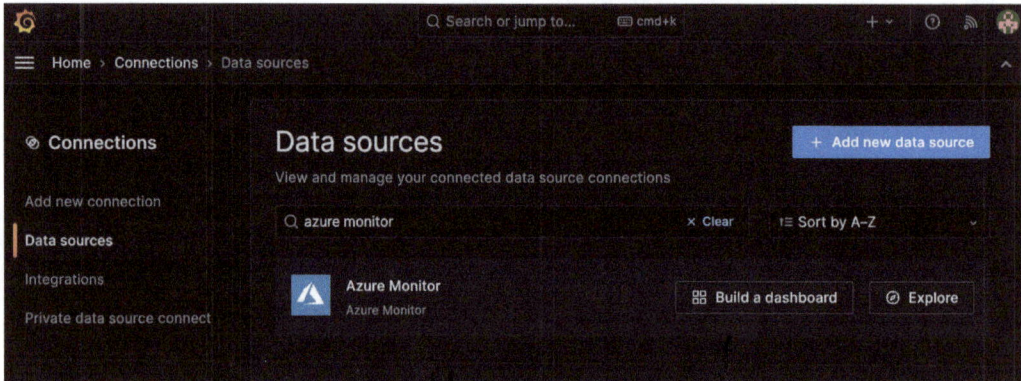

Figure 7.18 – Connection search results screen for Azure Monitor

Click on **Azure Monitor** to open the settings page. The configuration settings for **Azure Monitor** shown in the following screenshot walk you through the configuration, helping you to set up authentication using the Azure **Client Secret** configuration, and test the connection:

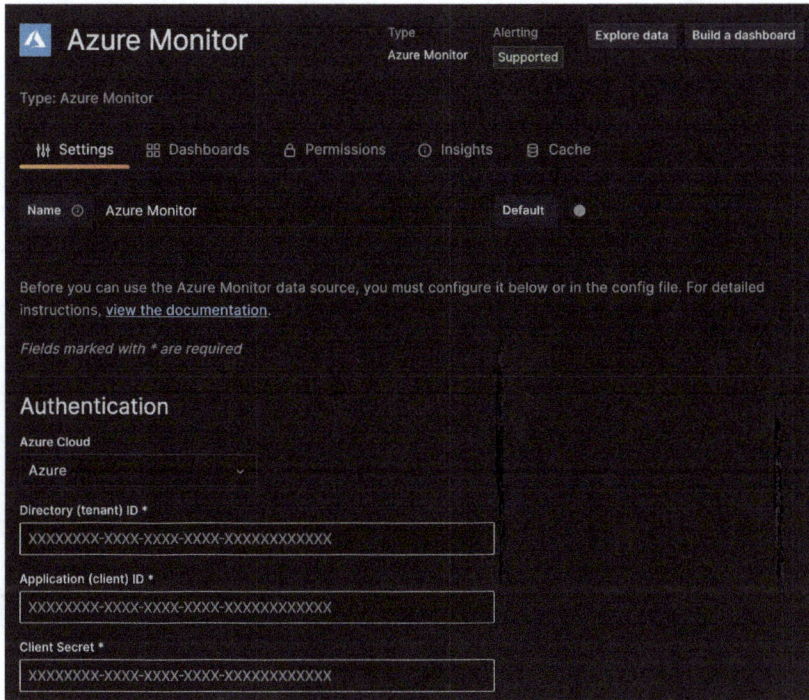

Figure 7.19 – Azure Monitor data source settings screen

Using the Azure Monitor query editor

The Azure Monitor data source comes with its own specialized query editor that can help you build queries for metrics and logs, Azure Resource Graph, and Application Insights traces.

The following Azure Monitor query editor screenshot shows four choices:

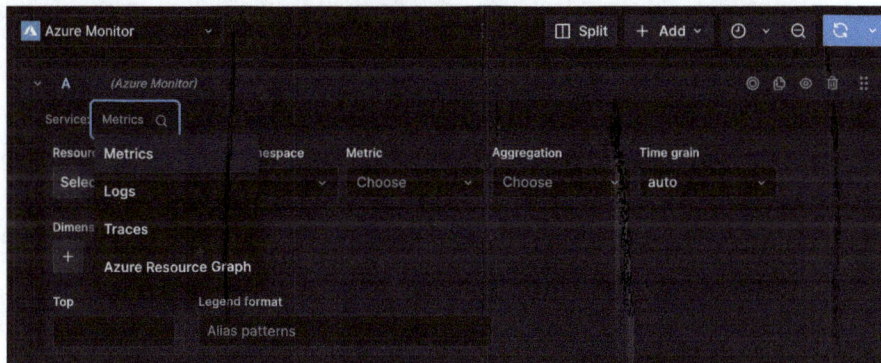

Figure 7.20 – Azure Monitor query editor selector

Let's look at these options in detail:

- **Metrics queries**: The Azure Monitor metrics queries collect numeric data from Azure-supported resources, which are listed here on the Microsoft Azure website: `https://learn.microsoft.com/en-us/azure/azure-monitor/monitor-reference`.

  The metrics store numeric data only, and in a specific structure that allows for near real-time detection of platform health, performance, and usage. The Azure Monitor metrics query builder is shown here:

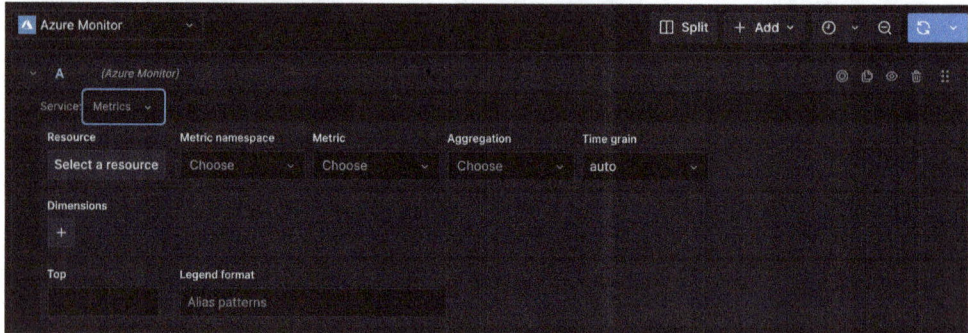

Figure 7.21 – Azure Monitor metrics query builder

- **Log queries**: The Azure Monitor logs queries collect and organize log data from Azure-supported resources. A variety of data types, each with their own defined structure, are accessible, and to access these, the **Kusto Query Language** (**KQL**) can be used. An overview of KQL can be found on the Microsoft Azure website here: `https://learn.microsoft.com/en-us/azure/data-explorer/kusto/query/`. The Azure Monitor logs query editor is shown in the following screenshot:

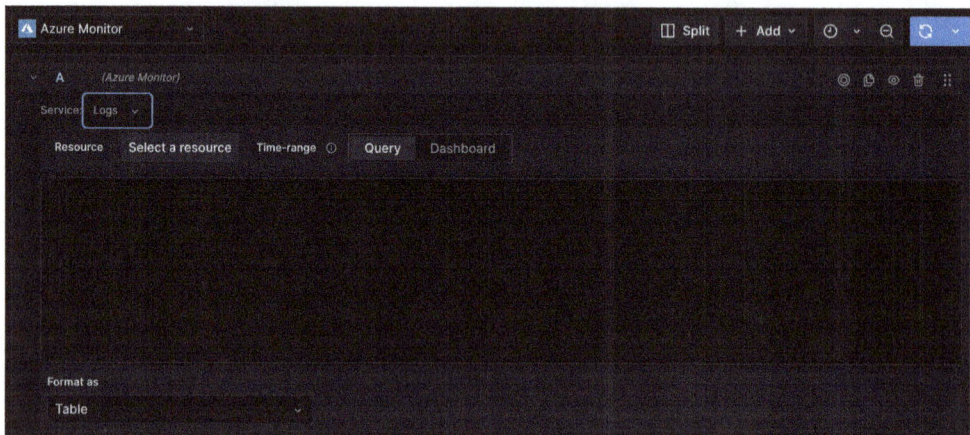

Figure 7.22 – Azure Monitor logs query editor

Azure Monitor logs can store a variety of data types, each with its own structure. For more details, you can refer to `https://learn.microsoft.com/en-us/azure/azure-monitor/monitor-reference`.

- **Traces queries**: The Azure Monitor traces queries can be regarded as Azure Application Insights under the hood. The Azure Application Insights service provides **application performance monitoring** (**APM**) features to its workloads. The Azure Monitor traces can be used to interrogate and visualize various metrics and trace data. The query editor looks like this:

Figure 7.23 – Azure Monitor traces query editor

- **Azure Resource Graph** (**ARG**): The ARG service extends the functionality of Azure Resource Manager by providing the ability to query across multiple Azure subscriptions in a scalable manner. This allows you to query Azure resources using the resource graph query language, making it ideal for querying and analyzing larger Azure cloud infrastructure deployments. Full documentation for the resource graph query language can be found at `https://learn.microsoft.com/en-us/azure/governance/resource-graph/samples/starter?tabs=azure-cli`.

The following example query shows all resources by name:

```
Resources | project name, type, location | order by name asc
```

Here's what the ARG query editor looks like:

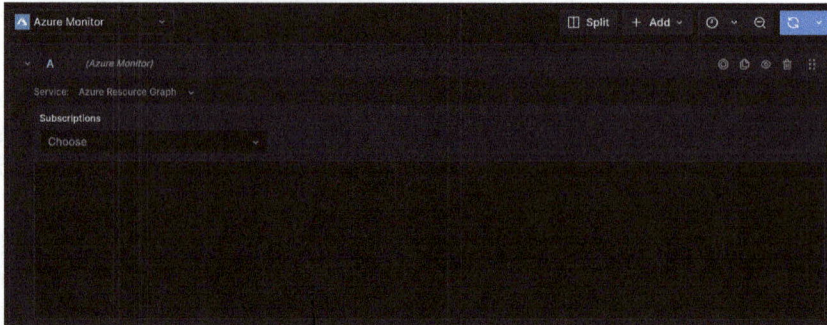

Figure 7.24 – Azure Resource Graph query editor

Using Azure Monitor dashboards

From the **Data source | Settings** screen, the **Dashboards** tab shows a set of pre-configured dashboards that will get you started with Azure Monitor. In the following screenshot, there is a list of the various Azure components that have dashboards designed for them; they include applications, SQL servers, and storage accounts:

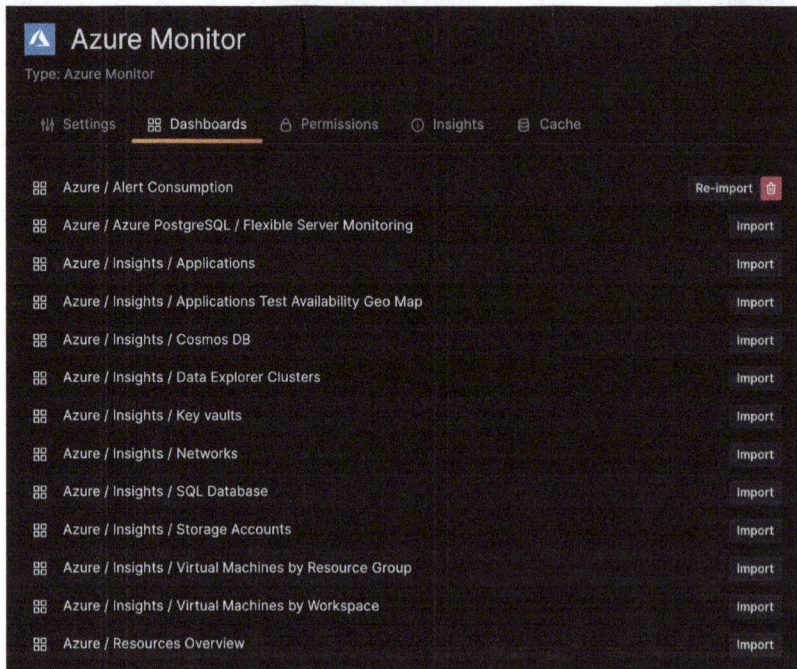

Figure 7.25 – Azure Monitor pre-built dashboards

Let's now review some of the best practices we have covered for each of the cloud infrastructure providers we have discussed in this chapter.

Best practices and approaches

In this chapter, we have provided an overview of several popular cloud infrastructures. Let's now discuss some of the best practices that should be considered when implementing observability on any application or system:

- **Performance**: The process of retrieving telemetry data can potentially incur a performance overhead. For example, with a remote Grafana data source, the telemetry data is fetched at query time over a great distance. This can introduce latency when compared to data stored closer to the Grafana query engine using one of the Grafana Cloud data sources, such as Loki, Mimir, and Tempo. Where performance is important and there is an option to ship telemetry into Grafana, that could be the best choice. Alternatively, several data sources have caching options to improve query speed; improvements in query speed can also be made using specific configurations. Take the time to understand your data and ensure you are using it in an optimal way.

- **Cost**: Alongside the increased network and storage costs of shipping data into Grafana Cloud, there can also be costs when querying a cloud provider API. It is important to understand where charges are raised. This ensures that they are factored in when you're designing your observability solution for the specific platform that your systems utilize.

- **Constraints**: In general, infrastructure platforms come configured with constraints in place to protect the system. Sometimes these are soft limits that can be relaxed after careful consideration, but they may be hard limits. Before committing to a solution for a specific platform, understand your requirements and the volume of data or query transactions expected. You can compare these to any documented limits, your API key use, or your network egress volumes, therefore validating that the system will support your needs.

- **Security**: For most of the configuration options we discussed in this chapter, we identified how they can be set up to generate separation of concerns. Having separate data sources or other controls on the data being queried or ingested will allow you to improve your security posture based on the underlying data and use case.

> **Important note**
> As this book was nearing publication, Grafana Labs released **private data source connect** (PDC), which gives administrators the ability to connect to any network-secured data source, regardless of where it is hosted. We have not covered this topic, but it is likely to be of interest to readers.

We will now wrap this chapter up with a summary and set the stage for the next chapter.

Summary

In this chapter, we have looked at various common cloud infrastructure providers, starting with Kubernetes, and we presented examples that can be used with the demo project provided alongside this book. We then looked at the big three cloud providers, AWS, GCP, and Azure. We presented an overview of the connection options and how to get started with the pre-built dashboards and the data explorer. Lastly, we covered some of the best practices that need to be considered with all observability integrations.

In the next chapter, we move on from getting telemetry data into Grafana and on to the visualization of that data using dashboards. This is where the fun starts!

Part 3: Grafana in Practice

In practice, Grafana is used to understand the current system state and take appropriate actions to give customers the best results. This part will cover the wide variety of activities that may be needed to complete that goal and what should be considered along the way. You will learn how to present your data while considering the requirements of your audience. You will explore how to build a world-class incident management process. You will also learn approaches to automating and architecting your observability platform.

This part has the following chapters:

- *Chapter 8, Displaying Data with Dashboards*
- *Chapter 9, Managing Incidents Using Alerts*
- *Chapter 10, Automation with Infrastructure as Code*
- *Chapter 11, Architecting an Observability Platform*

8

Displaying Data with Dashboards

Now that we have explored the topics covered in the earlier chapters together, we have a good understanding of the many ways to retrieve accessible data from Grafana. We will now create our first **dashboard** together and look at different ways to visualize our data. We will then explore concepts and techniques that help you communicate effectively with your dashboards and reduce cognitive load for your viewers. We will look at a few resources for inspiration out in the community to inspire you with your dashboards. Then, to finish, we will share guidance around managing your dashboard artifacts as they grow to help you get organized.

In this chapter, we're going to cover the following main topics:

- Creating your first dashboard

- Developing your dashboard further

- Using visualizations in Grafana

- Developing a dashboard purpose

- Advanced dashboard techniques

- Managing and organizing dashboards

- Case study – an overall system view

Technical requirements

In this chapter, you will work with the Grafana user interface using the Grafana Cloud instance and the demo you set up in *Chapter 3*. Also, to ensure you have all the data sources and settings available, be sure to have implemented the updates in *Chapter 6* to your stack. Let's dive in and create our first dashboard.

Creating your first dashboard

Dashboards are a medium to communicate important and sometimes urgent information to consumers. Before we delve into design techniques and best practices, let's get familiar with the Grafana user interface to start with a simple dashboard. The demo application we have worked with throughout this book has a frontend store; let's look at the requests we get for its shopping cart.

To create your first panel and dashboard, follow these steps:

1. In your Grafana instance, select **Explore** in the menu.

2. Choose your Prometheus data source; this will be labeled as **grafanacloud-<team>-prom (default)**.

3. In the **Metric** dropdown, choose **app_frontend_requests_total** (the metric count of requests to the frontend app), and in the **Label filters** dropdown, choose **target = /api/cart** as the filter (to restrict the results to the cart API only), and then click on **Run query**. You should see data like this:

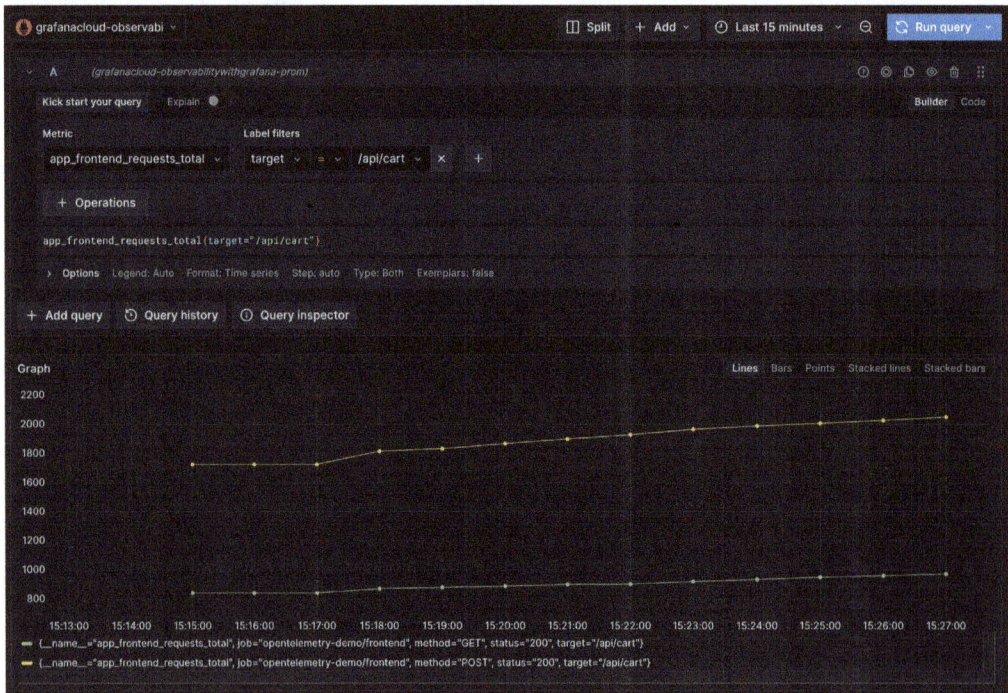

Figure 8.1 – App frontend requests

We now have a representation of the time series data in a graph.

4. To add this to a dashboard, select the **Add** drop-down menu button at the top of the **Explore** window.

5.   Choose **Add to dashboard**:

Figure 8.2 – The Add to dashboard menu

6.   From the pop-up screen, ensure **New dashboard** is selected, and then click **Open dashboard**:

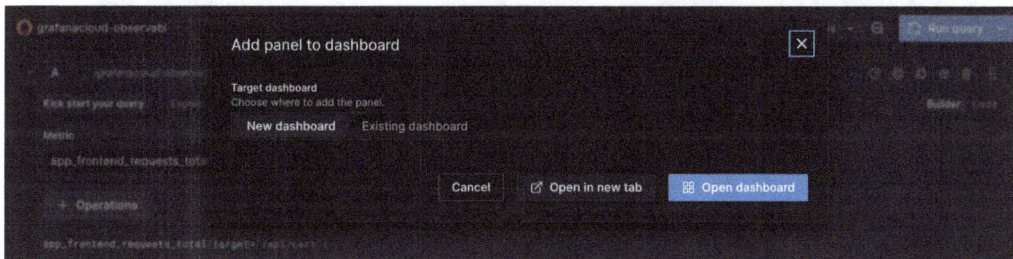

Figure 8.3 – Add panel to dashboard

You will now have your new dashboard on screen with its first panel. As you can see, there are a few problems – the legend is difficult to read, the data does not make much sense, and the panel title does not relate to the data:

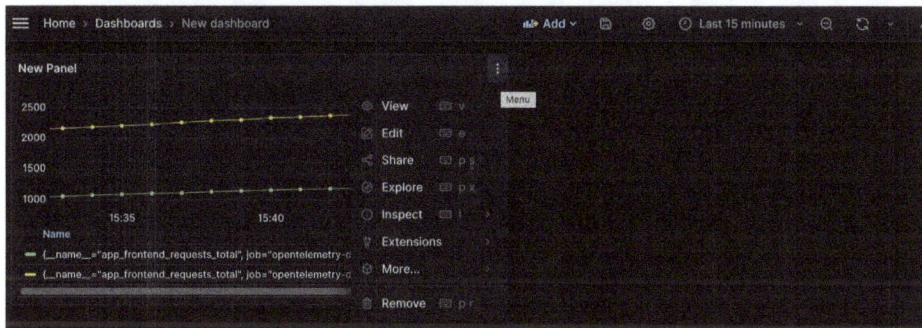

Figure 8.4 – The New dashboard view

7. Let's make a few changes to improve this a little. From the triple-dots drop-down menu, select **Edit**. The panel editor will appear; you should see something like this:

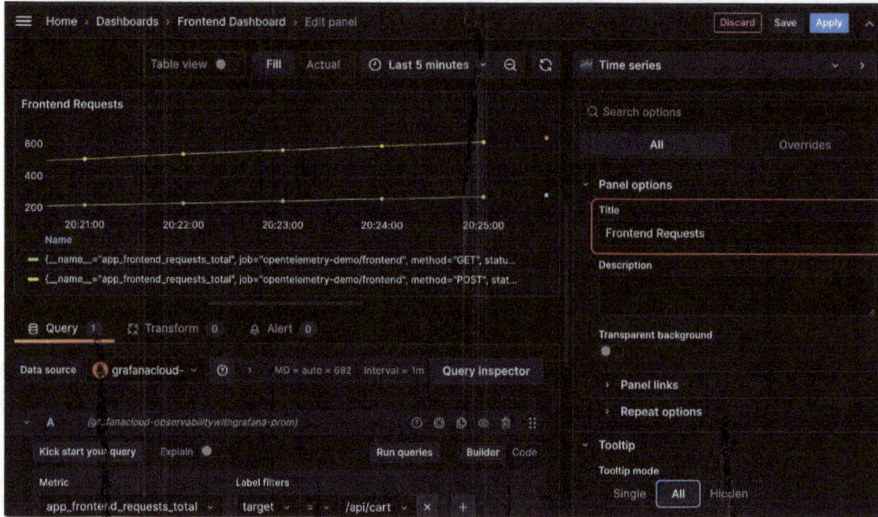

Figure 8.5 – The panel editor

8. Give the panel a meaningful name. We've entered `Frontend Requests` in the **Title** field:

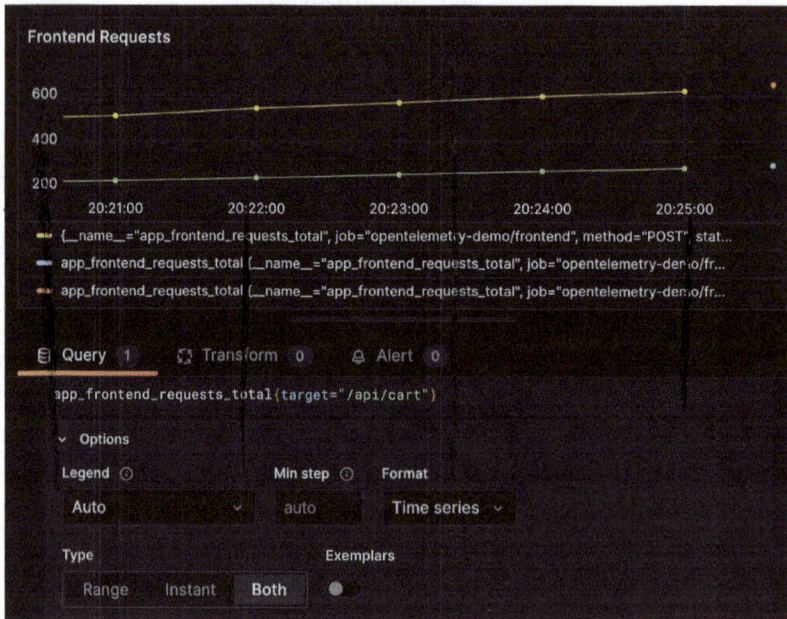

Figure 8.6 – Query options

9. The graph, by default, will show data from both **Range** and **Instant** queries, which can look odd on a time series visualization, as you will see a timeline and then standalone single values. Change **Type** to **Range** only.

10. Click the **Apply** button to update the panel and return to the dashboard:

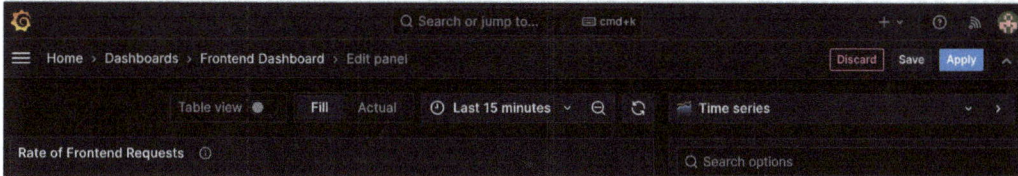

Figure 8.7 – The panel Apply button

11. You will now see the dashboard panel name applied to the metrics shown. Click the Save icon on the dashboard menu bar:

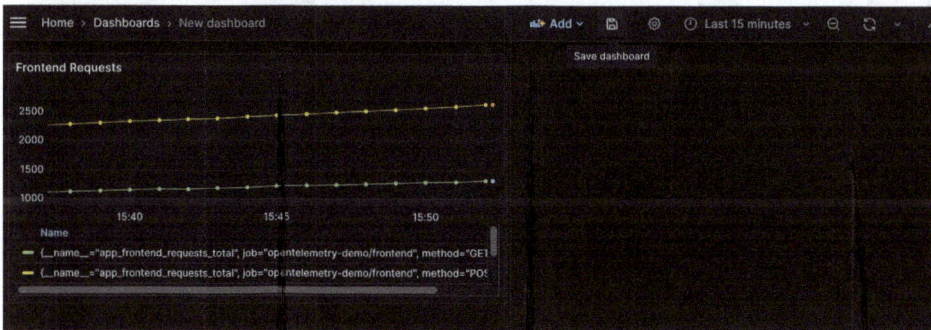

Figure 8.8 – Saving the dashboard

12. Give your new dashboard a meaningful name that reflects its purpose; we have a **Frontend Requests** panel, so let's call it `Frontend Dashboard`.

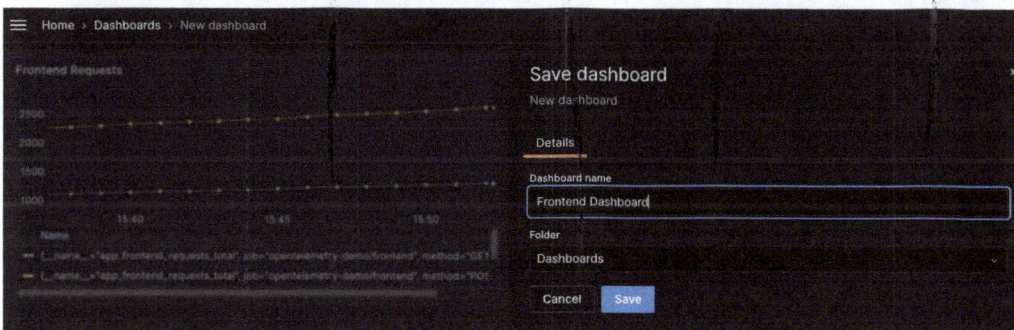

Figure 8.9 – The Save dashboard panel

We now have our first dashboard, but it is very simple, with only one panel. It does not effectively communicate much, and the number of frontend requests alone does not give an indication of what is happening to our online store. Let's explore what else we can do to improve this.

Developing your dashboard further

Let's now develop our dashboard further and improve the information we share. Approaching your dashboard as a work in progress that you can iterate over as you develop the message it is communicating is key to getting the right message across. In this section, we'll use an example where there's an increasing count of requests to the frontend app, which does not really tell us much. Showing the rate of those requests makes more sense, as we will be able to see the speed at which they arrive in our app. Let's make that change now:

1. From the panel's triple-dots drop-down menu, select **Edit** and you will see the following screen:

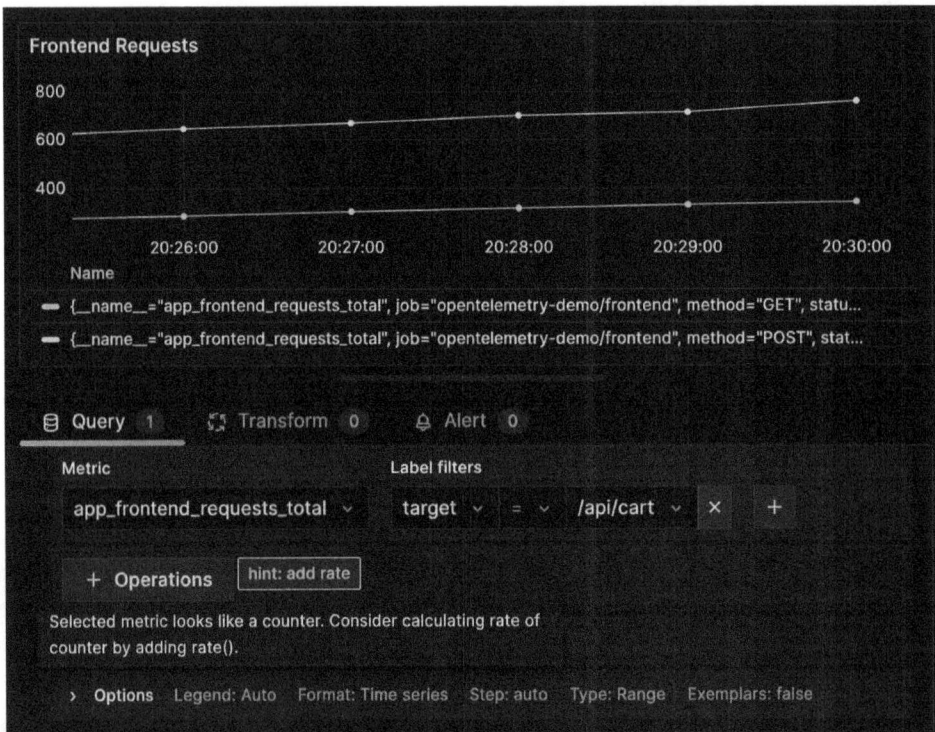

Figure 8.10 – The panel editor

2. Back in the panel editor, we can see that the query build gives us a hint to add a rate; this is highlighted with a blue box around it. Click on this hint to add the suggested aggregation to the existing query (if the hint does not appear, click the **Operations** button, and select **Rate** with a range of **$__rate_interval**):

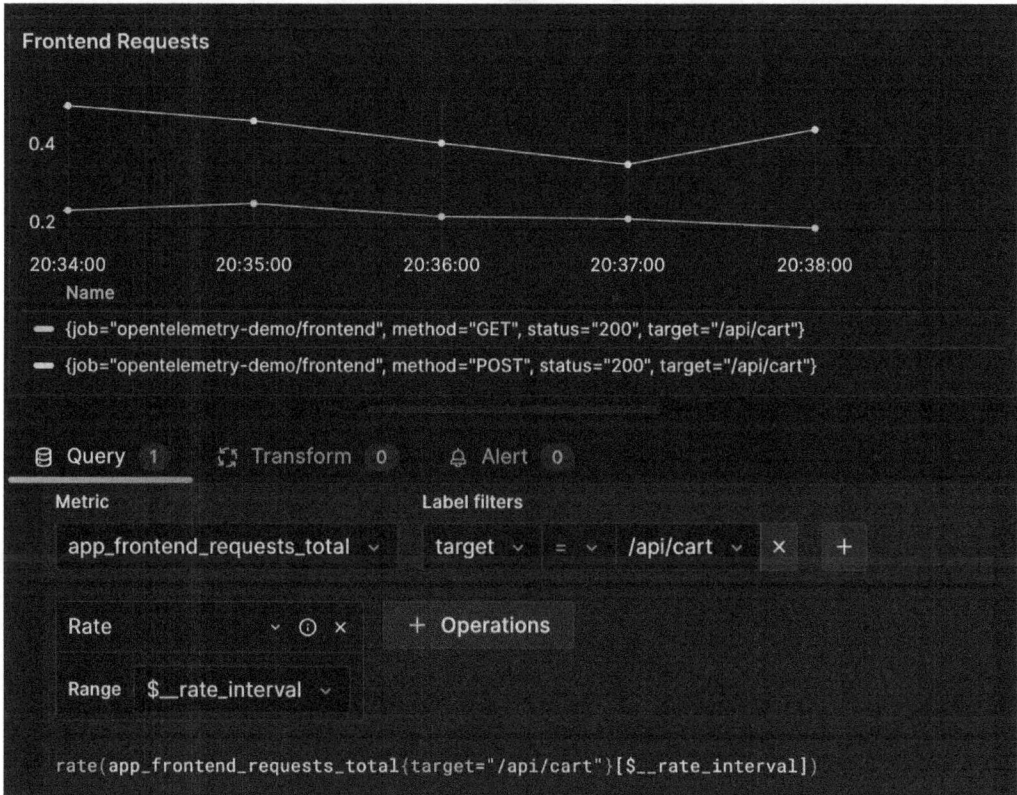

Frontend Requests

0.4

0.2

20:34:00   20:35:00   20:36:00   20:37:00   20:38:00

Name
- {job="opentelemetry-demo/frontend", method="GET", status="200", target="/api/cart"}
- {job="opentelemetry-demo/frontend", method="POST", status="200", target="/api/cart"}

🗄 Query 1   ⟲ Transform 0   🔔 Alert 0

Metric   Label filters

app_frontend_requests_total ⌄   target ⌄   = ⌄   /api/cart ⌄   ×   +

Rate   ⌄ ⓘ ×   + Operations

Range $__rate_interval ⌄

rate(app_frontend_requests_total{target="/api/cart"}[$__rate_interval])

Figure 8.11 – The Rate query

As you can see, the graph now shows how many frontend **GET** and **POST** requests per second come into our store. This is much more useful, but the legend is confusing, so let's clean that up.

3. Expand the **Options** section just below the PromQL query preview.

4. Replace the contents of the **Legend** field with the **Custom** label identifier ({{method}}), as shown in *Figure 8.12*; this will extract the values from the method label as the values for the graph's legend.

5. While we are here, let's also change the title to reflect we are now showing a rate, not a count, and add a meaningful description, as shown in the following screenshot:

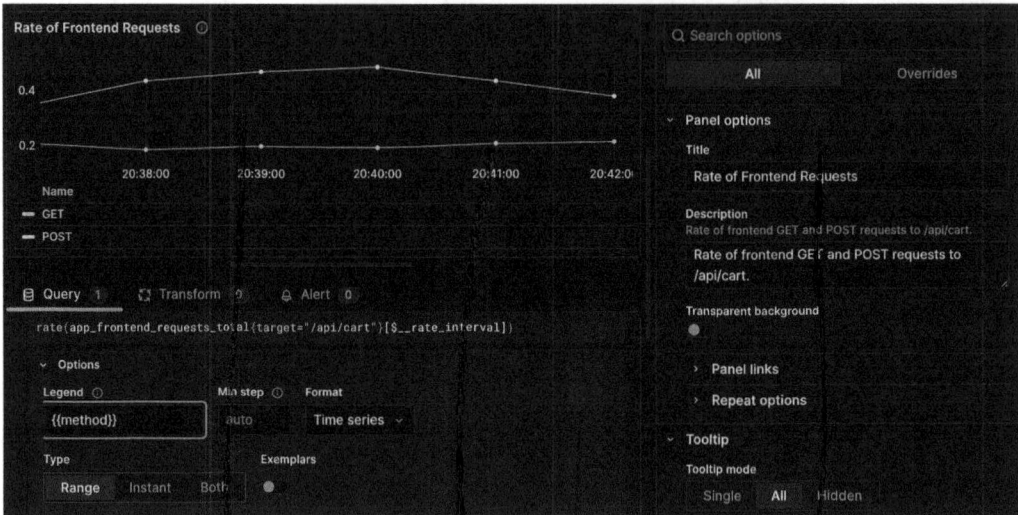

Figure 8.12 – Panel updates

6. Once again, click the **Apply** button to update the panel and return to the dashboard. You should see a dashboard like this:

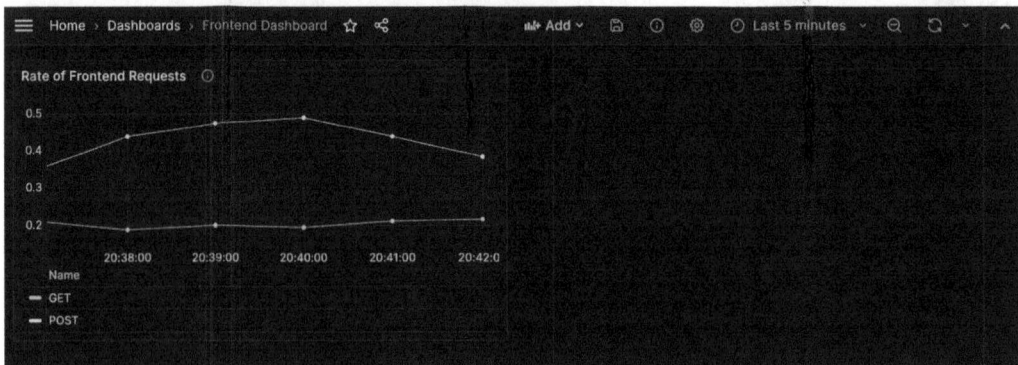

Figure 8.13 – Our first dashboard

As we can see in the preceding screenshot, we now have a meaningful presentation of the rate of **GET** and **POST** requests coming into the frontend API. The name of the dashboard and panel reflect their purpose well, and there is also a new **i** icon next to the panel, which will display the description you added when you hover over it.

Let's now look at the visualizations available in Grafana and how to use them.

Using visualizations in Grafana

Grafana has a large selection of visualizations to support your varying use cases and data formats. For example, for the counter and gauge metric types, you could use the **Stat** or **Gauge** visualization, and the histogram could use **Bar chart** or **Gauge**. Logs can be presented using the table visualization or by generating metrics from your logs in the time series chart. You can find the latest searchable list of plugins here on the Grafana website: `https://grafana.com/grafana/plugins/panel-plugins/`. A great place to try them out and get ideas you can use in your own dashboards is `https://play.grafana.org`, where you will find dashboard examples of all the panel visualizations along with a wide range of data types.

To change the panel visualization from within the panel editor, you can use the visualization selector, which is on the top right-hand side. Selecting it will display a searchable list, with a graphical representation of what the visualization looks like. This can be seen in the following screenshot:

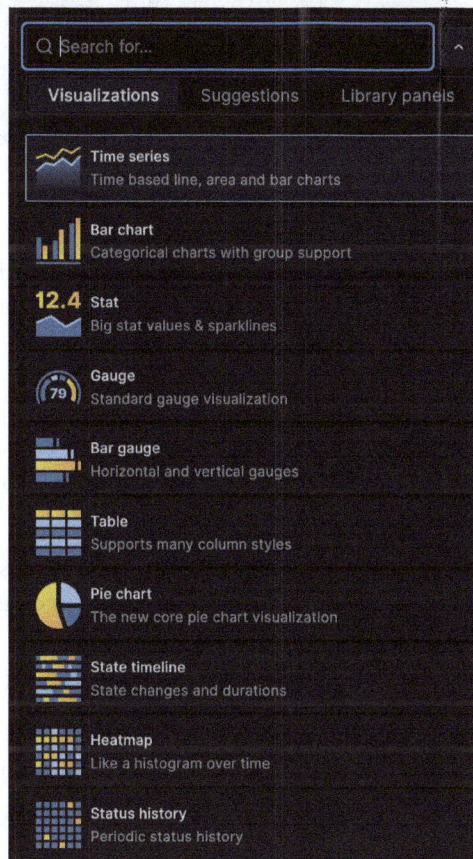

Figure 8.14 – Grafana visualizations

Let's try this with our panel on our first dashboard:

1. From the panel's triple-dots drop-down menu, select **Edit**.

2. Select **Stat** from the list of panel visualizations. You should see the dashboard panel change to look similar to the following:

Figure 8.15 – Stat visualisation

To better understand the options available, take some time to try out the different visualizations, or spend time on Grafana Play exploring the examples there. Once you know the visualizations available and the different panel configurations for each, it will become much easier for you to design your own dashboards.

A typical panel configuration will have the following items:

*   **A title and description**: While the title should help people understand the data presented, the description should complement or embellish the title.

*   **A legend configuration**: The legend should help interpret data presented with context and clarity. It can be formatted to support visually processing data and can interact with the view, filtering data to the selected item.

*   **A standard options setup**: This is used to control the options for the units, maximum and minimum values, decimals, display name, and color scheme.

*   **A panel-specific options setup**: Each visualization can have its own panel configurations.

*   **Thresholds**: These are used to control colors, backgrounds, or values displayed, based on whether a data threshold value is met or exceeded.

*   **Overrides**: Custom visualization settings can be applied to data to format and alter the presentation – for example, removing decimals and showing unit identifiers.

- **Value mappings**: These are similar to the overrides used to change the visual treatment of your data. They can be used to replace returned values with colors or values based on data matches (including ranges and regular expressions) to alter the visual presentation.

Spending time developing your familiarity with the configuration of some of the main visualization panels will give you the confidence to shape and present data for your users.

Let's now take a look at some design concepts that will help you present your data more effectively.

Developing a dashboard purpose

When creating a dashboard, it is very important that it has an objective. There are three key questions you can ask yourself to help you get the answers that identify those objectives:

- Who is the audience?
- What are their requirements?
- Where will the dashboard be viewed (on a big screen or a mobile phone)?

A dashboard must tell a story or answer a question, but we cannot do either of those things without answering the preceding questions. Let's refer to the personas introduced in *Chapter 1* to think about the type of people who need dashboards, what they want from them, and where they will see them:

	Persona	Requirements
	Diego the Developer	Insight into the traffic flowing through the systems I develop. This will inform how customers are using our product and help me make it better. Often, I need to investigate bugs and dig deeper into errors with much more detail.
	Ophelia the Operator	Visibility of problems as they happen, with clear indicators and colors to represent system states. Things need to be simple to understand at a glance. Dashboards will usually be on a large screen but need to be on my computer if I am investigating an issue.
	Steven the Service Manager	Overall, a system view that helps me see the wider picture. If there is a problem, I need the information to identify who can solve it. I will view this on my computer.

	Pelé the Product Manager	Clear and comprehensive metrics that show me how our products are used and with which devices. I will view this on my computer, and often, I want to export it to a spreadsheet for further analysis.
	Masha the Manager	Aggregated data is useful to me, showing trends with details on recovery rates and capacity so that I can plan ahead. I will view these on my computer, but periodic PDF reports via email help me.

Table 8.1 – Dashboard users and needs

Capturing end user requirements and delivering them with the dashboards you create will ensure value is taken from your observability platform and increase adoption. Once you have your user requirements and have analyzed your data (so that you understand what important information is there), you can start pulling your dashboards together. We say pulling together because it's important to remember you are creating something to answer a question, and that's going to take time. Do not be afraid to try out different styles, layouts, and visualizations as you work on the goal for each dashboard. Be sure to test under the different presentation requirements; if it's a large wall screen display or an emailed PDF, you want to ensure you have verified that it displays correctly.

Let's look at some techniques that will help with presentation.

Advanced dashboard techniques

To communicate effectively with our dashboards, we can use a layout and some technical features to up our game. Let's start looking at some of these in detail:

- **Layout tricks**: To help get your message across, you need to consider a few layout tricks that really make your message pop. These include the following:

  - Use a short top title in the **Overview** panel to make it clearer to read.

  - Add a description showing how to use the information and who it's for to give the users the best chance of understanding the data presented.

  - Call out key indicators using visualizations such as **Stat** with sparklines, and show them first at the top of your dashboard.

- Present panels in a grid to visually guide viewers through your dashboard and control their focus.

- Do not cram too many panels into a single dashboard; this makes it hard to digest. Instead, use datalinks to other dashboards and separate the data more.

- Use transparent panels to introduce visual spacing on your dashboard and help key data stand out.

- Use rows to group relevant panels and data together, creating a context for the data.

- Use the collapse feature of rows to hide certain data when the dashboard loads; this will also have the performance benefit of not running those queries until the row is expanded.

- Group similar visualizations to make the dashboard easier to read; this makes it easier to process the data presented.

- **Tech tips**: Grafana comes with some really useful features; here are a few tips to help you work smarter and speed up your development cycle:

  - Save dashboards often, or back them up by exporting the JSON.

  - Use useful change messages to help your future self when reviewing changes.

  - Remember that dashboard versions are saved, and these can be compared for changes if you need to debug a problem that has crept into your dashboard.

  - Create interactive and dynamic dashboards to engage viewers and deliver a message more effectively.

  - Use dashboard variables to get more use from your dashboard and help the viewer see data they are interested in; these are displayed as inputs at the top of the dashboard.

  - Avoid hardcoding values in panels and queries; your future self will appreciate this as the number of systems being monitored grows and new data is added that changes your dashboards.

  - Use relative time ranges as defaults to ensure the initial view is standardized – for example, set the from time to now-15m to show the last 15 minutes.

- Use transformations to present your data in more digestible formats. The **Transform** tab and available transforms are shown in the following screenshot:

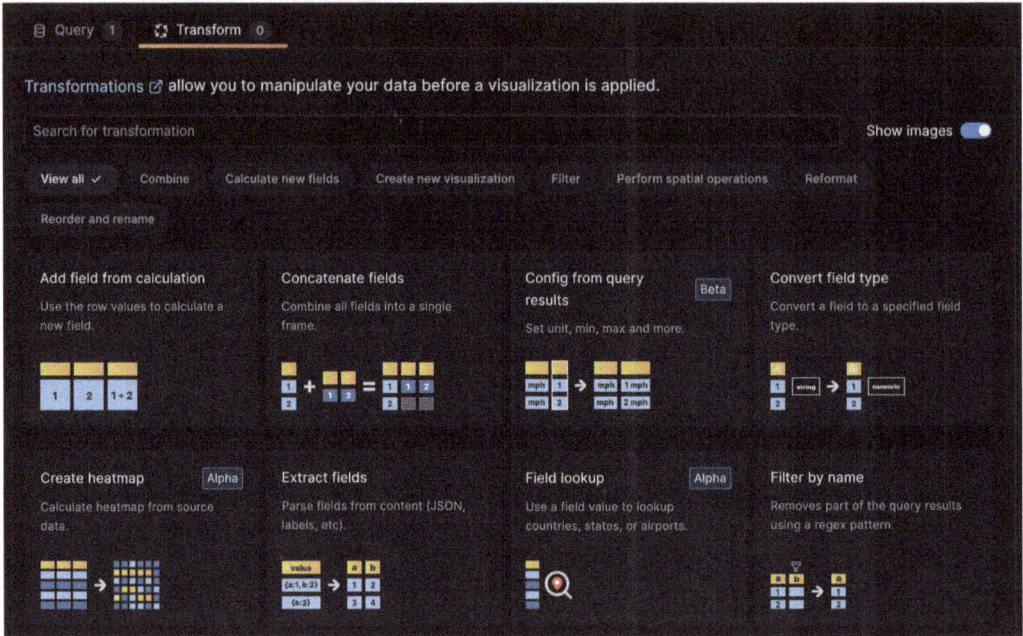

Figure 8.16 – The transformation selection screen

- Use annotations; mark points on a graph with rich events from other data sources or even manual annotations to communicate key events:

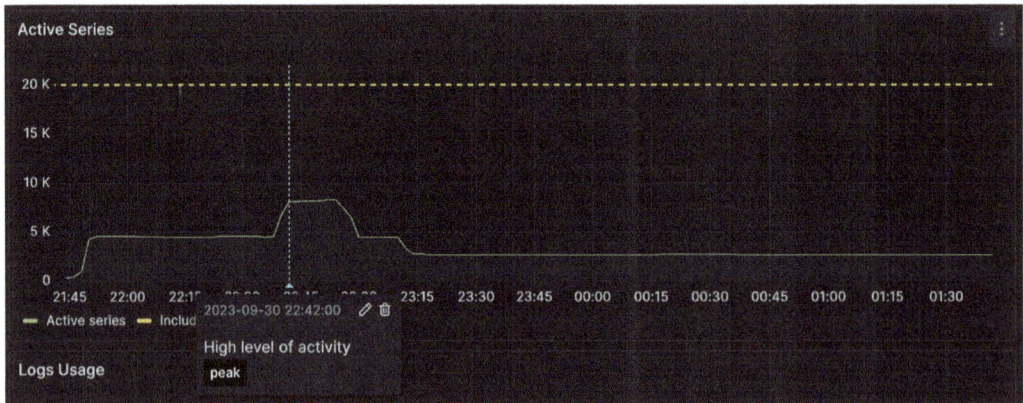

Figure 8.17 – Annotations

Let's now look at how we can develop good practices to maintain control over our dashboards.

Managing and organizing dashboards

As the list of dashboards and teams using them grows, you will need a way to manage them easier. Grafana provides folders, tags, and permissions that give us the capability to get organized. From the main **Dashboards** screen, you can create folders, move and delete dashboards, and star them, as you can see in the following screenshot:

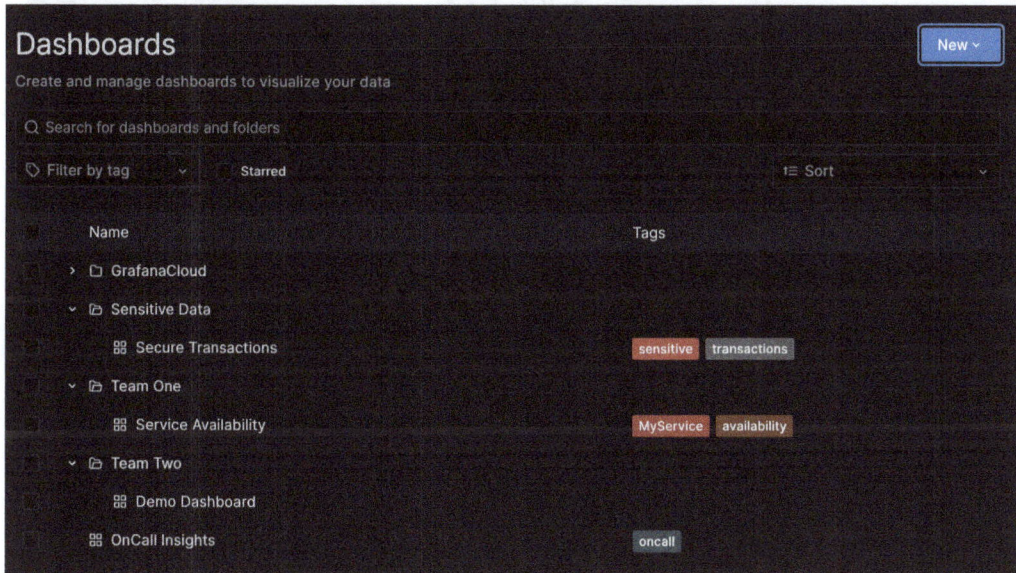

Figure 8.18 – The Dashboards screen

From the dashboard **Settings** screen, you can add or remove tags from individual dashboards, as shown in the following screenshot:

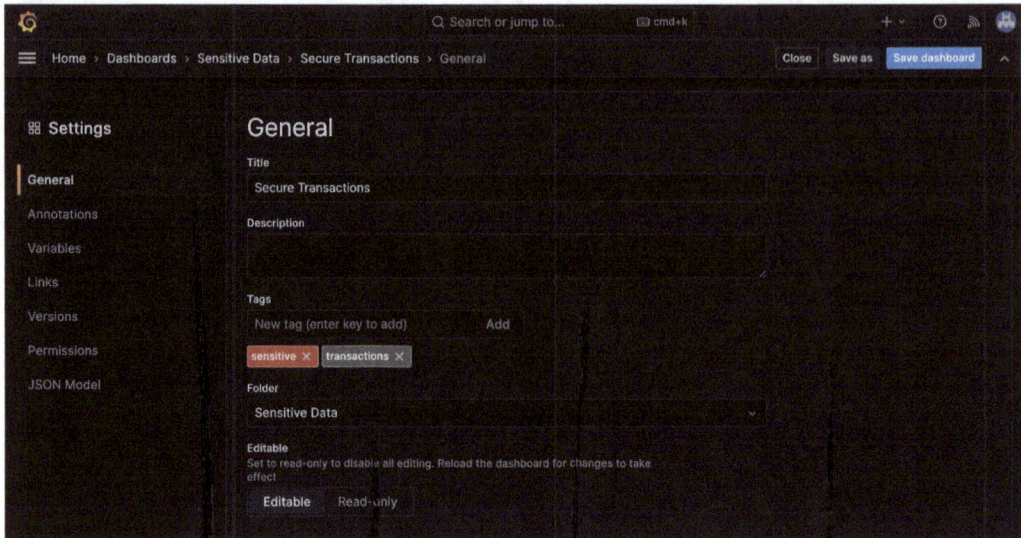

Figure 8.19 – Dashboard settings

Let's look at the available settings for folders, tags, and permissions:

- **Folders**:

  - Provide easy organization and grouping

  - Enhanced security by segregating permissions at the folder level

- **Tags**:

  - Useful when searching for related dashboards

  - Color-coded for visual recognition

- **Permissions**:
    - Control access to folders and their dashboards
    - Give viewer access only to those who do not need to edit
    - Hide folders on a *need-to-know* basis
    - Available for both individual users and teams

Now that you have an understanding of the concepts involved, let's look at a case study to develop a dashboard for one of our personas.

Case study – an overall system view

We will now walk through the process you would take to develop dashboards for *Steven* the Service Manager. If you recall from *Chapter 1*, he works in service delivery. He wants the organization's services to run as smoothly as possible.

We will start with a use case; in an earlier section in this chapter, *Developing a dashboard purpose*, we recorded the following requirement from Steven:

> *"Overall, a system view that helps me see the wider picture. If there is a problem,*
> *I need the information to identify who can solve it. I will view this on my*
> *computer."*

This can be broken down into the following:

- A top-level system view
- Problem indicators (to highlight issues)
- More detail when needed
- Assistance in identifying who can help

We then need to gather information from documentation and by having conversations.

We have been using the OpenTelemetry demo app throughout the book; fortunately, architecture diagrams are provided at `https://opentelemetry.io/docs/demo/architecture/`, which will help us understand the system better.

Let's look at the OpenTelemetry demo system architecture diagram:

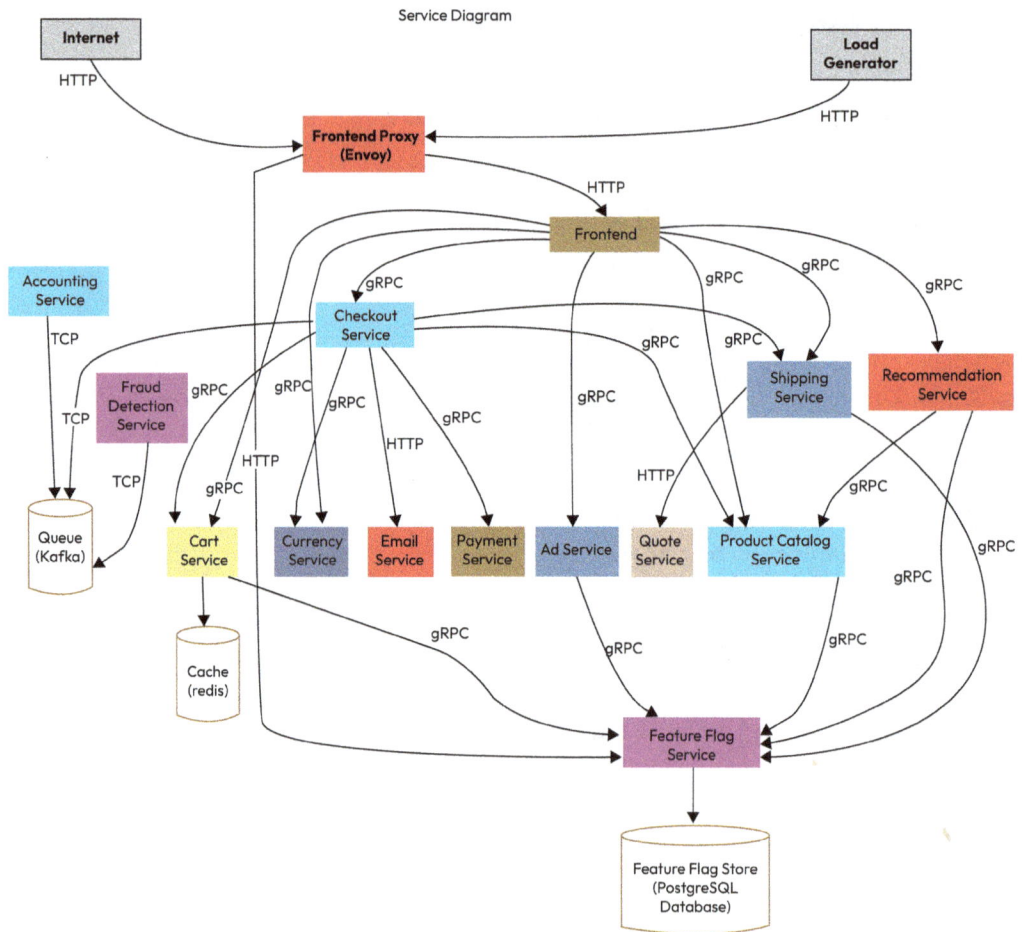

Figure 8.20 – OpenTelemetry demo system architecture

We will also look at what telemetry is being ingested for these applications. This could be done by exploring Grafana ourselves; we have system names now and application flows. Alternatively, we could talk to *Diego* the Developer and obtain his insight into the system.

If you are exploring by yourself, this is a good time to build a research dashboard. This is where you add panels as you explore and discover interesting data. We did this with our first dashboard earlier in this chapter, starting with frontend requests and then enhancing it to show the rate of frontend requests. Otherwise, start building up a scrapbook dashboard of data you may want to use; do not worry about layout or visualizations right now.

To help us choose useful metrics, we can use a methodology such as the **Four Golden Signals**, as identified in Google's *Site Reliability Engineering* handbook (https://sre.google/sre-book/monitoring-distributed-systems/#xref_monitoring_golden-signals). There are other popular methods available – for example, the **Utilization, Saturation, Errors** (**USE**) and **Rate, Errors, Duration** (**RED**) methods, which we will discuss in the next chapter.

Let's quickly take a look at the Four Golden Signals:

- **Latency**: The time taken to service a request
- **Traffic**: How much demand is placed on your system
- **Errors**: The rate of requests that fail
- **Saturation**: How full your system services are

When you have some key metric data ready for use, you can start shaping your dashboard. Remember that a dashboard should tell a story or answer a question; *Steven* has said he wants a top-level system view that highlights problems with access to more detailed information as and when needed. We can follow these steps:

1. Let's start with the first row of the dashboard, creating the first-look area. By choosing metrics detailing the flow of traffic through our system, we can use visualizations such as **Stats** (with a timeline) to stand out. Lead the viewer, *Steven,* from left to right, with the most important information first.

   Here's an example of **Stats** with timeline visualizations for the first-look area:

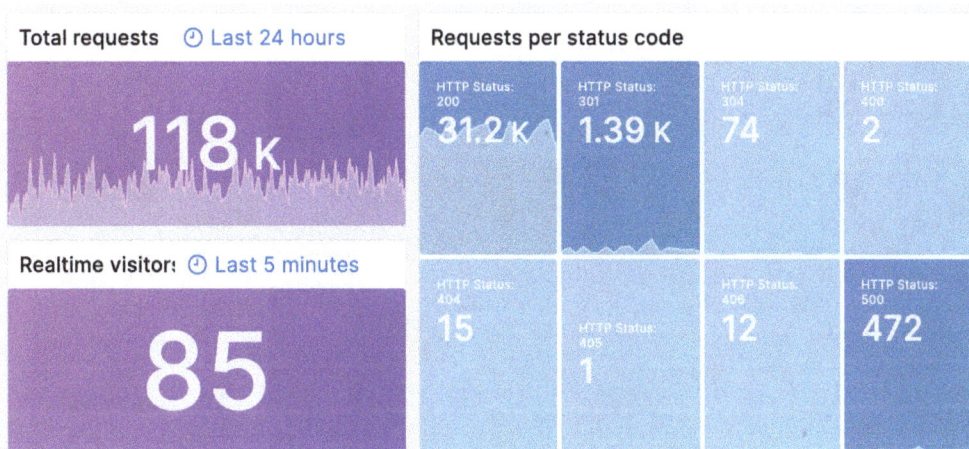

Figure 8.21 – A system overview

Other visualizations such as **Gauge** are good to represent the saturation of a system.

To ensure Steven's attention is drawn to the most important information first, we can enhance our panels with thresholds, overrides, and value mappings. In certain panels, we can even overlay further context as annotations.

2.　We then need to find a way to communicate more detailed information and decide this based on how much information we have to display. We have to ask two questions:

- Can we present the detailed information here on this dashboard itself (is there enough room)?

- Do we need to drill down into other, more focused dashboards?

For our use case, we will work on the *same* dashboard. To support *Steven* with the identification of who he needs to support in the event of an incident, we can do the following:

- We can logically group our visualizations in rows. Each row can be a team, and the first panel can be a text panel used to share contact information and provide details on how to engage the team. This means the cognitive load when using the dashboard is greatly reduced, helping *Steven* solve problems more quickly.

- We can also use panel descriptions to share useful information that helps viewers, including *Steven*, understand the data, and dashboard links to callout support pages will also help.

- To represent more detailed information better, we could use visualizations such as the table panel along with color coding to provide visual aids, helping to process the information quicker.

Here, we have an example with top-level insights and more detailed and directional (i.e., identifying the owning or supporting teams) panels and rows:

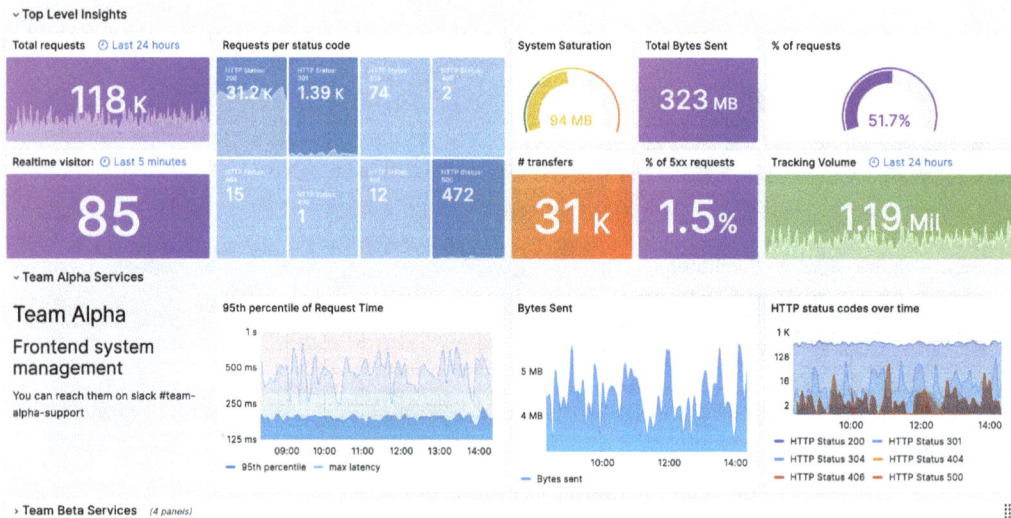

Figure 8.22 – A detailed dashboard

As you can see from this fictional example, building a useful dashboard requires several iterations. Much more than telemetry alone goes into the process, but the results mean *Steven* can do his job and have the best shot at delivering customer satisfaction.

Let's now close this chapter with a few words on how you can support your development with dashboards.

Summary

In this chapter, we explored the different components of a dashboard and looked at the different details we need to factor in. These techniques will take time to build, so approach your dashboard development in the same way, get something simple built early, and iterate on it. We talked about the visualizations available for panels and inspiration on websites such as `play.grafana.org`. Have some fun, and try out alternative methods to present your data and find a style you like. Finally, we looked at some tricks and tips to improve your dashboarding, wrapping up with some simple ideas to help you manage your dashboards.

In the next chapter, we will learn about **incident management** and the great tools available from Grafana that support it.

9

Managing Incidents Using Alerts

This chapter will explore the concepts of **incident management**. We will discuss how to build a world-class incident management process, which treats those responding to incidents humanely and avoids burnout. The chapter will establish the responsibilities for this, from the senior leadership teams to the engineers responding to the callout. It will introduce the important concepts of building an organization that can handle incidents and excel at providing customers with a stable experience. With the process established, we'll explain how to consider a service and pick critical measures that can be used to see the current service level, without being drowned out by noise.

This chapter will also explore the three tools available from Grafana for incident management. First, there's **Grafana Alerting**, which is used to monitor metrics and logs for failures and trigger notifications to responding teams. Then, there's **Grafana OnCall**, which expands on the features of Alerting with a dedicated mobile app for alerts, team schedules, and the ability to receive alerts from any third-party application that can send webhook data. OnCall lets you centralize all alerts for visibility and route them to the right response team. Finally, there's **Grafana Incident**, which provides an easy-to-use incident tracking tool that keeps all highlighted information ready for your post-incident activities, helping your organization focus on improvements that prevent incidents.

In this chapter, we're going to cover the following main topics:

- Being alerted versus being alarmed – how to build great incident management
- Writing great alerts using **service-level indicators** (SLIs) and **service-level objectives** (SLOs)
- Grafana Alerting
- Grafana OnCall
- Grafana Incident

Technical requirements

In this chapter, we will go into technical details, aimed at readers such as *Ophelia Operator*, *Diego Developer*, and *Steven Service* (who represent operators, developers, and service delivery professionals, as introduced in *Chapter 1*), and which may be of interest to readers such as *Masha Manager* (in leadership), to understand what is possible with Grafana's tools and how they support established incident management.

Being alerted versus being alarmed

Incident management is a common process used in many areas, from physical incidents such as fire and medical emergencies to computer security or service failure. While we may not handle life-threatening incidents in the computing world, the stress caused by bad incident management processes can be very significant, from anxiety and depression to complete burnout, and it can increase the chance of heart attacks and strokes. Our aim in this section is to explain how observability and Grafana's tools fit into an incident management strategy, and how to use them to reduce the impact on your teams, the duration of incidents, and the frequency of incidents. We will explore the details of these concepts and the tools available in Grafana to support them further throughout the chapter.

There are a lot of great public resources available on the topic of incident management; here are some for you to explore if you wish:

- *Emergency response and recovery* (https://www.gov.uk/government/publications/emergency-response-and-recovery): This is guidance given to the emergency services in the UK. While most of you will not be handling incidents that involve risks to life or property, this document is a fantastic read for anyone looking to understand how to make incidents as easy as possible to handle.

- *Atlassian Incident Handbook* (https://www.atlassian.com/incident-management/handbook): The *Atlassian Incident Handbook* is a great place to start when writing or reviewing an incident management process.

- *What is incident management?* (https://www.servicenow.com/uk/products/itsm/what-is-incident-management.html): Similar to the *Atlassian Incident Handbook*, the ServiceNow incident management guide is a great place to start when writing or reviewing an incident management process.

- *Google Site Reliability Engineering* (https://sre.google/sre-book/managing-incidents/ and https://sre.google/workbook/incident-response/): Google's *Site Reliability Engineering* books are packed with helpful information. Most organizations will not be running services at the same scale as Google, but these give a clear view of creating a highly scalable incident management process.

This may seem like an oversimplification of incident management, but we will group these concepts into *before*, *during*, and *after* an incident.

Before an incident

As the mantra says, *"Hope for the best, and prepare for the worst."* Knowing how you will respond to an incident before it happens is crucial to responding effectively to an incident when it happens. In this section, we will discuss various aspects that need to be in place before an incident occurs.

Roles and responsibilities

Incidents are messy, quickly evolving situations, and they are no place to be trying to figure out who is doing what. Roles and responsibilities for your organization's incident response must be clearly documented and understood by everyone who *may* be called on to respond to an incident. It is not advisable to reinvent the wheel for incident management; there are several frameworks available, including the following:

- **Information Technology Infrastructure Library (ITIL)** incident management
- **Site Reliability Engineering (SRE)** incident management
- The **National Institute of Standards and Technology (NIST)** incident response framework
- The **SysAdmin, Audit, Network, and Security (SANS)** incident response framework

There are some key roles that appear in all of these:

- **Commander**: This is the person who has the authority to make decisions. These are some key features of this role:

  - The range of decisions will be different for each incident, but the commander needs to be able to call in the correct people, sign off on communications to customers, and handle internal communications with senior leadership

  - This is also the person who has overall control of an incident

  - All other roles report to this person

- **Communications**: This is the person who is responsible for internal and external communications. These are some key features of the communications role:

  - Communicating effectively during an incident is vital for success, and this person is responsible for managing that

  - Internal and third-party communications are their responsibility

  - Customer-facing communication is also their responsibility

- **Technical leader**: This person is vital for directing the many technical people who may be involved in an incident. Some key features of technical leadership during an incident are as follows:

  - When multiple people from multiple teams are investigating a problem, it's important for one person to make the technical calls

  - The technical leader needs to know who in the technical teams is looking at what and when to expect updates on findings

The roles outlined in these frameworks are very focused at the **operational** (*bronze*) level. The UK emergency services have documented a very effective command structure, **gold–silver–bronze**, which outlines responsibilities for **tactical** (*silver*) and **strategic** (*gold*) levels. It is valuable to be clear about the responsibilities of executive or senior leaders and their subordinates *before* any incidents are handled. This ensures everyone involved in an actual incident knows how to bring in the correct leaders when needed. What we mean here is that it's better to have a plan that can handle a major incident that causes serious harm to an organization than to realize you need one during the incident. Let's explore these levels in greater detail:

- **Gold – strategic responsibilities**: Gold teams are made up of senior managers or the C-suite. Members of the gold team should always keep their focus on the strategic level and not get drawn into making tactical decisions. The main responsibilities of gold teams are as follows:

  - Set, review, and communicate the incident management strategy

  - Define whether any resources or specialist skills are needed

  - Handle the media strategy

  - Consider the legal issues that may arise from any incidents

  - Report to shareholders where appropriate

  - Approve the silver team's tactical plans before they are used

  - Lead the de-brief or postmortem after an incident

- **Silver – tactical responsibilities**: Silver teams are composed of managers from different departments. They provide tactical leadership for bronze teams, make decisions on how to implement the strategic vision set out by gold teams, and, during incidents, act as the conduit for information to flow between gold and bronze teams. Silver teams are responsible for the following:

  - Set, review, and communicate the tactical plan for incidents up and down the chain of command

  - Document the incident management procedures

  - Capture how communication with customers should be handled

  - Choose the appropriate tools to manage incidents

- Understand which teams will be meeting which strategic objectives

- Address any resource needs in critical teams

- Update the gold team with any relevant information during an incident

- **Bronze – operational responsibilities**: The bronze team is the team that is responsible for responding to an incident, from the initial alert to the conclusion of the post-incident process. The bronze team's responsibilities include the following:

  - Taking operational control of the incident

  - Informing the silver team when an incident is declared

  - Understanding the cause of an incident

  - Making decisions on how to resolve an incident

  - Communicating internally and externally within the tactical plan

  - Completing post-incident reports and meetings to address any ongoing issues

Cutting noise to improve the signal during incidents

It is impossible to know how an incident will occur and what the root cause will be. When they do occur, getting the right information quickly and communicating effectively are two very important factors in recovering from the incident as quickly as possible.

The first place to reduce noise is in **observability systems**, which makes it easier to identify important signals. Doing this requires knowledge of a service, which is why it is best to do this before an incident occurs. The following practices can help engineers such as *Diego* share the detailed domain knowledge of their application with the wide experience of systems from engineers such as *Ophelia* and *Steven*:

- Identifying and documenting critical SLIs

- Using distributed traces, so all calls can be seen in service graphs

- Writing log messages that are easy to understand

- Writing logs that handle failures without producing lots of messages

A lot of observability tools offer some form of **AIOps**; these effectively offer a tool that watches the standard flow of data and highlights times when something deviates from the previously seen data. This should not be treated as a reason for not identifying critical signals, as these tools do not replace specific domain knowledge in our experience.

The second form of noise that can occur is the inappropriate use of **communication channels**. Many of us will have seen messages such as *Is the site down right now?* in a Slack or Teams channel, dedicated to notifying us of incidents when they become a problem on a user's computer. Noise and lack of signal can occur for customers as well, either repeatedly notifying customers of every minor blip, or more likely, not informing customers when an incident occurs. Getting these communications correct is a core part of an incident management strategy. Common practices include the following:

- Giving a dedicated group of people authority to declare an incident and its severity
- Automating the communication response to an incident when declared – for example, updating a customer-visible status page and posting in a dedicated internal communications channel
- Setting up protected internal channels of communication for incident teams (*bronze*)
- Setting up protected channels of communication between incident teams and senior leadership (*bronze* to *silver* and *silver* to *gold*)
- Pre-writing status messages so that incident teams only select the most appropriate message for most consumers

Supporting tools

The adage that a bad workman always blames their tools is very appropriate here; there is no *perfect* tool and what works for one organization may not work for another, and there are many tools on the market. This is a list of capabilities that we believe all organizations should consider as part of their incident management strategy:

- Alert notification – sending a page to the person on call. Also, consider mobile apps for out-of-hours notification.
- Process automation.
- On-call rota management.
- Integration with other systems.
- Automatically capturing internal communications during an incident.
- Running practice or drill incidents.
- Escalation processes.
- Customer-facing communication during an incident.

Now that you have a lot of knowledge on preparing for the worst, let's look at these plans in action during an incident.

During an incident

Incidents happen, as much as we would prefer that they didn't. In this section, we'll discuss some key tasks that help make the process of resolving incidents as painless as possible.

Identifying the incident

There are many failure modes that can be seen in computer systems, from immediate lights-out outages through cascading failures to intermittent failures. Having clear information to say when a service is behaving *as expected* is vital to identifying issues. It is the responsibility of domain experts such as *Diego* or *Ophelia* to write this information into software services, or ensure they are provided by systems from third parties, such as compute, storage, or network services. There are some common ways of capturing this information, including **white-box monitoring techniques** and **black-box monitoring techniques**.

White-box monitoring is the practice of monitoring a system that you have access to. This practice helps identify whether a service is healthy, with detailed information of the state of the service. Here are some ways of presenting metrics that are commonly used in this process:

- **Rate, Errors, Duration (RED)**: This is a way of measuring services that are driven by requests. *Rate* is a measure of the volume of requests the service handles in a period. *Errors* is the number of requests that are encountering errors. *Duration* is the distribution of request durations, and it's common to represent this as a set of percentiles or a histogram.

  With these three signals, we can quickly compare the current state with a "normal" state, checking whether any service has a higher or lower number of requests hitting it, whether it has a higher than usual number of errors, or whether the duration of the requests are longer or shorter. With that knowledge, the incident response team can identify services that need further investigation.

  RED is a great system to use for any service that responds to requests, such as a web server.

- **Utilization, Saturation, Errors (USE)**: USE and RED are complementary to each other; RED looks at the requests to a service, and USE looks at the internal state of the service. *Utilization* measures the number of resources the service is using to process work (we'll talk about resources shortly). *Saturation* is the amount of work that the service cannot process due to a lack of resources. *Errors* is the number of errors that are being produced. We've used the term *resources* here; these will be different in each service, and identifying them is an area for domain expertise. Common resources would be CPU and RAM availability, network or disk I/O, or even the number of threads available in an application.

  USE is best-suited to model a service that offers a resource, such as a storage system or Kubernetes cluster.

- **Golden signals**: Golden signals were introduced in the Google SRE book, and they overlap very strongly with RED and USE. The golden signals are **latency**, **traffic**, **errors**, and **saturation**.

  Errors and *saturation* are the same as described in RED and USE. *Traffic* is the measure of requests per second, so is equivalent to the rate from RED. *Latency* is the time it takes to service a request; this is like duration from RED. However, latency also captures whether a request is successful or not. This signal allows for differentiation between situations where a duration may be lower because the error is returned very quickly and the more challenging scenario of a service taking a long time to give an error.

- **Core web vitals**: The previous views of services were driven by the backend systems. Core web vitals are a set of metrics that are gathered from an end user's browser, usually using a **Real User Monitoring** (**RUM**) agent such as Grafana Faro, which is embedded into web applications to collect data. This set of metrics is very focused on the end user experience for web applications.

  The current core web vitals include the following:

  - **Largest Contentful Paint (LCP)**: LCP measures the loading performance of a web page; it is a measure of when the largest element on a page is rendered. Historically similar metrics such as First Contentful Paint, First Meaningful Paint, Load, DOMContentLoaded, and SpeedIndex have been used; LCP is the current recommendation from Google's web.dev team.

  - **First Input Delay (FID)**: FID measures the interactivity of a page; it is a measure from when a user first interacts with a page to when the browser can process the event handlers in response.

  - **Cumulative Layout Shift (CLS)**: CLS measures how frequently the content rendered on a page changes position because another element was rendered.

In contrast to white-box monitoring, black-box monitoring treats a service as an unknown and checks whether it is behaving as an end user would see it. There are broadly two categories of black-box monitoring:

- **Synthetic monitoring** uses a configurable service to simulate connections and user actions. Prometheus offers the black-box exporter, which can connect to endpoints using HTTP, HTTPS, DNS, TCP, **Internet Control Message Protocol** (**ICMP** – often called `ping`), and **Google Remote Procedure Call** (**gRPC**). Other services on the market can also simulate critical user journeys if more granular external monitoring is needed. Where **service-level agreement** (**SLA**) adherence is a contractual obligation of an organization, using a third-party synthetic tool is a very easy way of proving that the SLA was (or was not) adhered to. This tool is offered as a managed service as part of Grafana Cloud as **Synthetic Monitoring**. All functionality except gRPC calls are supported.

- **RUM** uses an agent embedded in frontend code to collect data from real end users using a service. While RUM offers much broader functionality than just black-box monitoring, it can also be used to provide the initial alert based on the actual experience of end users.

Black-box monitoring does come with a risk of false positives. While white-box monitoring only covers items that are under the control of the organization such as internal networks, cloud-provided services, and so on, black-box monitoring must cover areas outside of control such as external internet provider networks or DNS services.

By using common groups of signals and clearly defining these critical SLIs for each service, organizations can effectively transfer the knowledge of a service's health from domain experts to others in the organization. Detailed how-to guides commonly known as **runbooks**, which detail responses to situations, also transfer this knowledge. Finally, a robust post-incident process effectively allows organizations to step away from a *hero culture*. A hero culture is when a small group of individuals keep things working by responding to every incident, often at the expense of their health. A mature organization is one that moves from the chaos of frequent incidents to one that gives highly motivated individuals, and the organization as a whole, space to grow.

Once our monitoring has notified us that there is something wrong, the next questions are *who* to bring into the incident and *when*.

Escalating an incident

If an incident can be resolved quickly with only the on-call person being involved, this is ideal. Unfortunately, some incidents need to be escalated, whether that is because someone with more specialized knowledge is needed, or because the scale of the incident is too large for one person to handle. Each organization is different, and a clear escalation policy needs to be part of the tactical plan for incidents. An escalation policy should give clear guidance and answer these questions:

- Who should be notified when an issue is identified by an automated system?

  - Does this change during in-hours and out-of-hours?

  - Does this change depending on severity?

- Who should be notified if the first responder isn't available?

- Who should take over if a first responder can't resolve an issue alone?

- What criteria are used to make that decision?

  - The duration of the incident?

  - The severity level of the incident?

  - The time of day of the incident?

- How should the handover of an incident occur?

- What happens if there are multiple incidents at one time?

With this guidance in place, it is the responsibility of leadership (*Masha*) to make sure everyone who is on call knows the policy. This is especially important for junior engineers who may feel that they need to avoid disturbing more senior colleagues. There is also a responsibility to regularly audit on-call schedules and ensure engineers on call are protected from overwork and burnout from the schedule.

Communication

During an incident, communication is critical. We can split communication into three broad strands:

- **Incident team communication**: Most organizations use some combination of in-person or video meeting rooms and chat tools. There are a few considerations that should be taken to make this communication easy during an incident:

  - What is the primary communication channel to tell someone to join an incident?

    - A chat channel, phone call/SMS, or mobile app (e.g., Grafana OnCall or PagerDuty)

  - Is there a primary conference bridge video meeting?

    - Make sure the details are included in any alerts sent on the primary communication channels

  - What is the expected response time for people called into an incident?

    - This has an impact on the time to recovery

    - This also has an impact on the health and well-being of people regularly called into incidents, which should be monitored

  - How are team communications recorded for post-incident review?

    - Peoples' recollection will become fuzzy as the time between an incident and the post-incident review increases. Capturing communications is a good way of managing this to ensure that the post-incident process is as accurate as possible. The tools that can be used to make this process as simple as possible during an incident are very valuable to the entire incident management process.

    - There are tools and processes to assist here. Grafana Incident will track a timeline of events from tools such as Slack. The communications and technical leads of an incident should also be responsible for regular status updates, which should be made with a post-incident review in mind.

- **Internal communication**: The communications lead of a major incident is responsible for internal communication. For most channels, the tactical plan for incidents should specify a frequency of updates. This communication typically does not need to go into a lot of detail; even saying, *"We are still investigating the issue and working to resolve it"* is better than being silent. This communication is equally important during incidents of internal tooling as well.

The communications lead should keep senior stakeholders such as the gold and silver teams informed of the current state in more detail. As silver teams will typically include technical and managerial leadership for the products that may be the cause of the incident, this channel is especially important for escalating and bringing in experts where needed. Following the same primary communication channel for incident notification is the best practice – that is, if escalation is a case of notifying the correct on-call rota, then incidents can be resolved more quickly. However, this does come with the cost of placing more engineers on call.

- **Customer communication**: This is likely the most important because when incidents affect external customers, it is important to get a message out quickly to reassure an organization's customers that you are on top of the issue and working on restoring service. There are a whole host of options for communication with customers:

  - Status pages, either dedicated and separate from the organization's service or embedded into it

  - Email notifications

  - Social media

  - SMS notifications

  - Messages on any customer ticket management portal

  During an incident, the communications lead should have a selection of pre-generated and approved messages for customers, with little need to modify the message. This helps keep the tone and feel of the messages correct, even if the communications lead has just been woken up at 3 a.m. It is also useful to have a message that informs customers there may be a problem but you are investigating; this is ideal for situations where you've been alerted to a situation but you're unsure whether there is an actual impact on customers.

With all these things in place, you will have put your organization in a great place to resolve incidents quickly and let the team go back to bed. After the incident, the arguably harder pieces of work will begin, such as understanding why the incident happened and communicating how the organization is going to fix any underlying causes. Let's explore how to approach this.

After an incident

Incidents happen to big organizations and small organizations; even organizations that have been meticulous in avoiding incidents will experience them. The most important thing from any incident is for the organization, as a whole, to learn about the vulnerabilities in a system or the gaps in processes that led to the incident.

When incidents happen, it can be natural to look for a person or a department to blame; this human tendency is in direct contradiction to an organization's best interest. Blame leads to burdensome procedures, a lack of innovation, and ultimately, to the organization's stagnation, as staff stop being honest and seek to ensure they are not blamed, demoted, or even fired.

Blameless postmortems are a space for an honest and objective examination of what happened, with the goal of understanding the true root cause(s). Good intentions from all staff and departments must be assumed. It is critical to understand that the goal of a blameless postmortem is not to remove accountability from an individual or team but to ensure that accountability is not accompanied by the fear of reprimands, job loss, or public shaming. The important aspects of a blameless postmortem include the following:

- Open communication where mistakes are accepted as a part of life

- Encouraging honesty and the acceptance of failure

- Sharing detailed information on the timeline of events, which should be supported by the logs of internal communication and systems during the incident

- Making decisions and seeking approval for improvements

There are many guides that detail these processes in much greater detail than we have gone into here; our goal is to introduce those of you who fit the personas to the broad topic of incident response. We will now discuss in more detail how the practices of observability and the tools of Grafana can help build part of a great incident response plan. We will start this by looking more deeply at SLIs, as well as SLOs.

Writing great alerts using SLIs and SLOs

An **SLI** is a measurement that is used to indicate a current service level. An example could be the number of errors over a 15-minute period.

It is best practice to keep the number of SLIs small; three to five SLIs for a service is a good rule of thumb to follow. This reduces confusion and allows teams to focus on what is critical for their service. SLIs can also be thought of as a **fractal** concept; while a service team can have indicators for a component of a larger system, the system can also be tracked by a small number of SLIs – for example, the number of services that are failing their SLOs. By keeping the number of SLIs tracked relatively small, the potential for spurious alerts is reduced, and the impact of continuously monitoring services is kept small. This means more services can be monitored without scaling the tools used and increasing operating costs.

The patterns we discussed earlier of RED, USE, golden signals, and core web vitals are good SLIs to consider when deciding what to track. These are not the only measures that can be used as SLIs, but they have good adoption in the industry and are well understood. Teams should think hard about whether they need to use something different.

By agreeing on SLIs and objectives, the process of writing a great alert becomes much simpler, as the person implementing the alert only needs to consider how to translate the business description of the SLI (the number of errors in a 15-minute period) into a query, using LogQL or PromQL, and then create a threshold based on the set objective. Alerts written this way will also be easy to understand.

Another important concept is the SLO, which refers to an internally agreed-upon target that is considered acceptable for an SLI. An example would be no more than 3% of requests resulting in errors during a 10-minute period.

While not directly related to incident alerting, there is the concept of **error budgets**, which are strongly related to a good SLO setting. An error budget is an SLO that measures the meeting of other SLOs. When the error budget is exceeded, this is a good indicator that a service is unstable in some way, and this should serve as a trigger for focusing the team's energy on remediating this. Conversely, if the error budget is high, this should give the team the space to experiment, or even take the service down in a planned way. This can be a great opportunity to expose issues that could be catastrophic if they were seen in an unplanned outage. This topic is discussed in much greater detail in publications that discuss SRE.

An SLA is an agreement made with clients or users on what is acceptable for the service. These can be legal agreements. These are often made up of multiple SLIs and SLOs. An example would be an uptime of 99.9%. Setting objectives with SLOs helps an organization keep easily within their legally agreed SLAs, which is a great way to ensure that the SLAs are very rarely breached and customers feel they can trust your organization.

We've explored the theory of a good incident response strategy and the choices to make, from the team level to the organization level, to support it. Let's now take a look at the three tools Grafana offers to support incident response, Grafana Alerting, OnCall, and Incident.

Grafana Alerting

Grafana Alerting is the first of three major components of Grafana's **Incident Response and Management** (**IRM**) toolset. Grafana Alerting itself comes with no additional licensing costs, and it is an ideal solution for smaller organizations while forming the foundation of the IRM tools in larger organizations.

The IRM features can be accessed from the main Grafana menu under the **Alerts & IRM** tab:

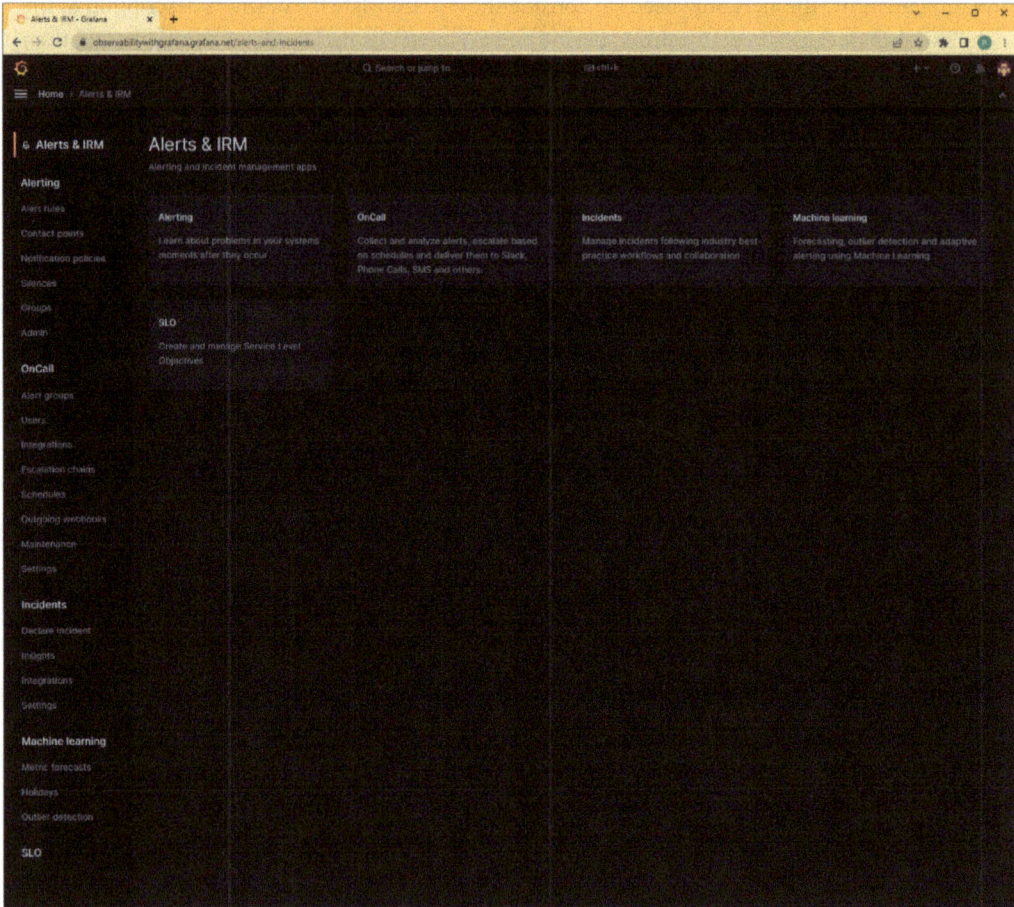

Figure 9.1 – The Grafana Alerts & IRM main screen

Grafana Alerting continuously evaluates user-created **alert rules** for alert-worthy states, following predefined steps to send messages to the chosen notification channel.

Next, we will see how to set up alert rules, get your contact points and notification policies right, silence alerts when needed, and set up teams and team members.

Alert rules

The main configuration screen for Grafana Alerting is the **Alert Rules** screen. This allows you to set up new rules and see the current state of existing rules. Setting up a rule should feel very familiar after learning LogQL and PromQL in *Chapters 4* and *5,* respectively.

Setting up an alert rule requires several items to be configured; let's walk through those now:

1. **Set the alert rule name and define the query and alert condition**: First, a query is created and named; in our example, we have used the following query for a period of 10 minutes:

    ```
    (sum(app_frontend_requests_total{status=~ " 5.. "})/sum(app_
    frontend_requests_total))*100
    ```

    This will calculate the percentage of all requests in the period that were completed, with a status code of 5xx, which is the SLI. In the **Expressions** section, we can then set how we want to reduce the time series returned; in this case, we want the last event, but if we were looking perhaps at errors per endpoint, we could get the maximum value, so we see the highest error rate from any endpoint. The **Threshold** value lets us set when the alert is triggered or not – in this case, when the error percentage is above 3, which is the SLO. The following screenshot shows the screen where these items can be filled in:

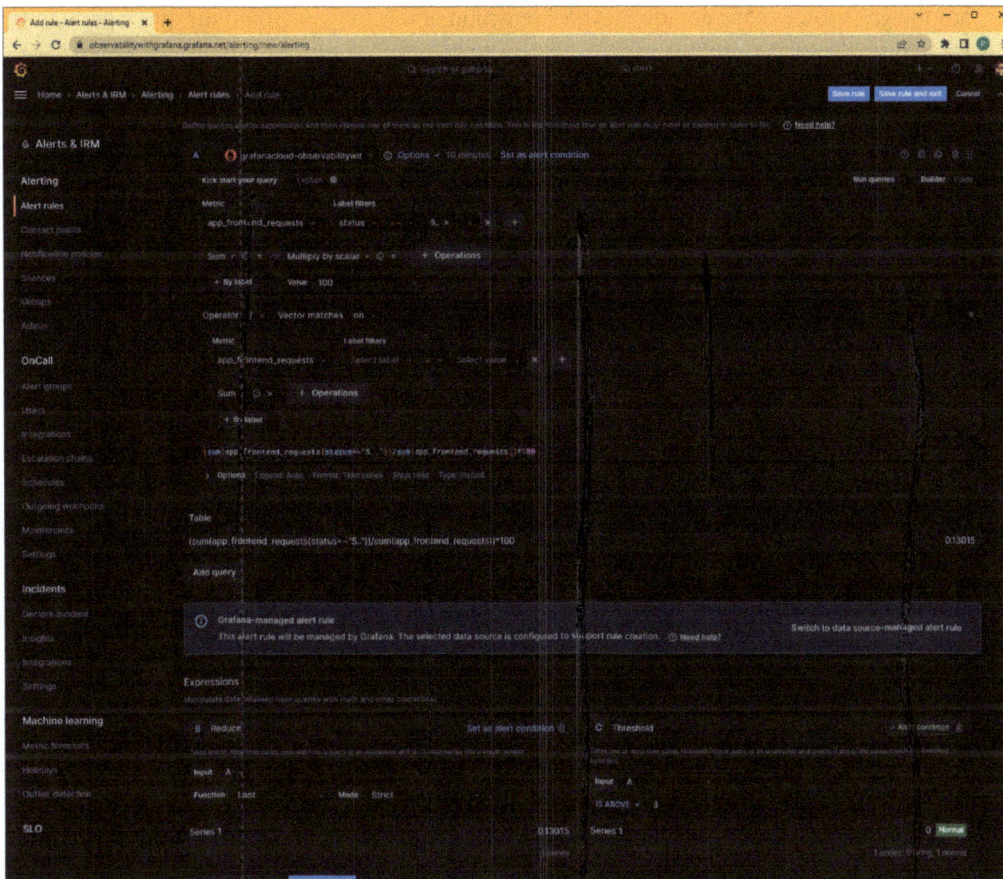

Figure 9.2 – Creating an alert rule

2. **Set the alert evaluation behavior**: With our SLI and SLO set, we now need to decide how we want Grafana to evaluate its next action. Grafana has three states an alert rule can be in – **normal**, **pending**, and **firing**. The **alert evaluation behavior** manages how an alert rule group will transition between states. An evaluation group will sequentially evaluate each rule in the group with the same evaluation period. In our example, we have created a frontend group, which would contain the RED metrics from the frontend service. As our frontend service is business-critical, the evaluation period is set to 1 minute, and our pending period is set to 5 minutes. With these settings, if our error percentage goes over 3, our rule will enter the pending state within 1 minute, and our alert will trigger within 5 minutes if the state persists. This can be seen in the next screenshot.

It is very tempting to set these values as low as possible (10 seconds); however, this can have unintended consequences. Running the queries every minute would result in 1,440 queries per day, while every 10 seconds would give 8,640 queries per day. While compute power is relatively cheap, this still increases the resources required by Grafana by six and probably offers little advantage. Another consideration is the interaction of this frequency and the query period. If we evaluated every five minutes but our query only looked at the last minute, we would have minutes that were not evaluated, which could disguise legitimate intermittent errors. Let's look at the lower part of the screen to manage an alert rule:

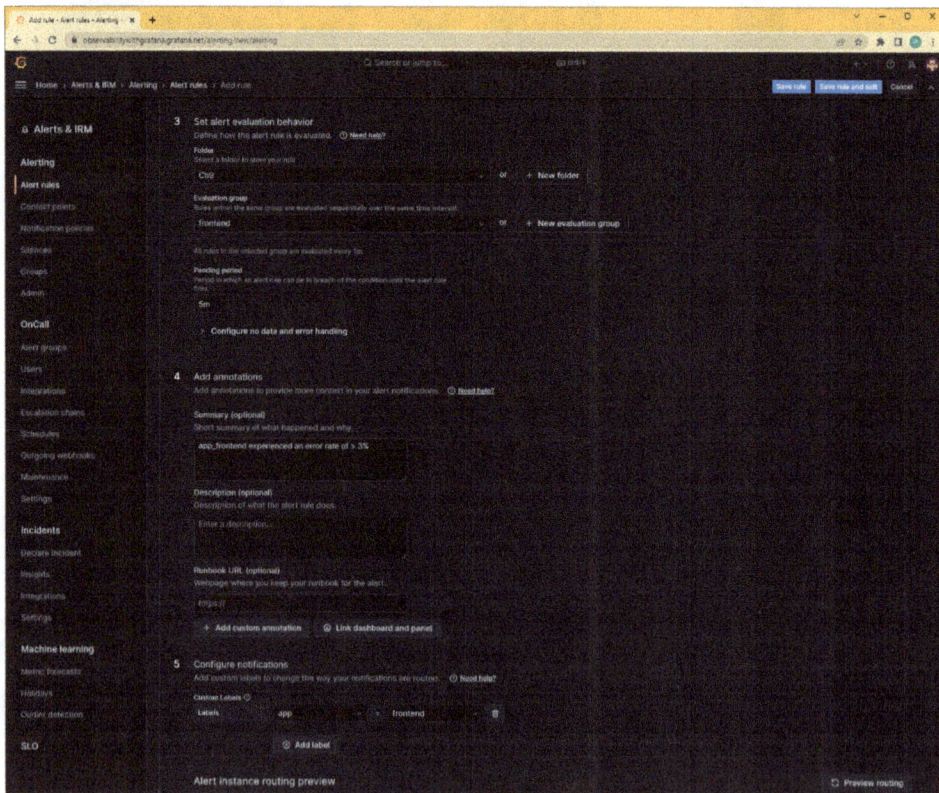

Figure 9.3 – Managing an alert rule

3. **Add annotations**: Annotations are used to add information that will be sent to the contact point when an alert triggers. It is good practice to include easy-to-read summaries that capture key information, such as which service is involved and what the problem is. The description should give more detail if needed, and it is best practice to include a runbook URL and dashboard link. These should guide a responder to quickly understand the problem and give information on remedial steps they can follow to quickly restore a service.

4. **Configure notifications**: The **Configure notifications** section provides space to add labels. These can be used to manage alert routing, which we will cover when we discuss notification policies. Clicking the **Preview Routing** button will give information on how an alert will be routed with its current configuration.

The next few tabs in the **Alerting** menu allow us to configure other aspects of Grafana Alerting. These are smaller items than the main alert rules. Let's take a look at them now.

Contact points, notification policies, and silences

Contact points are configured by Grafana admins. They consist of a name and one or more integrations. It's typical for a contact point to be a team responsible for addressing an issue. With some of the integrations available, such as webhooks and Kafka message queues, it is relatively easy to establish more complex contacts that go beyond just alerting. In more complex environments, Grafana offers the option of creating notification templates on the **Contact points** page. These can be used to standardize the message structure across multiple contact points and integrations. Grafana provides a great guide to setting up custom notifications here: `https://grafana.com/docs/grafana/latest/alerting/manage-notifications/template-notifications/`.

Notification policies allow you to connect alerts triggered from an alert rule to a contact point. These policies can be nested, and they can also have mute timings applied to silence notifications during off hours. A very valuable feature of notification policies is the ability to match notifications on a label from Loki or Prometheus/Mimir sources. This effectively gives you the ability to use a label from the source telemetry and route notifications to the team responsible for that service; an example of this could be routing to teams on the `service_name` label from the services in the OpenTelemetry Demo. Rules at the same level of nesting can also continue matching, meaning the service team and a central operations team can both be notified.

Silences are defined periods when no notifications will be created. These can be used to manage maintenance periods when using Grafana Alerting.

Groups and admin

The **Groups** section shows grouped alerts. Grafana will also display alerts here for data sources that have defined alerts but are not sending data.

The admin page provides the configuration for Alertmanager in JSON format; this allows administrators to save a configuration and transfer it to other instances, or to use it as a configuration backup.

Grafana OnCall

Grafana OnCall is the second major component of IRM and expands on Grafana Alerting, by adding capabilities to do the following:

- Consume alert notifications from many external monitoring systems

- Specify alert groupings to reduce noise during incidents

- Specify when alert groups should send notifications

- Define on-call rotations and escalation paths

- Expand notification channels from what is offered in Grafana Alerting:

  - Create and update tickets in ServiceNow, Jira, and Zendesk

  - Notify the current on-call individual directly

- Provide a mobile application for engineers to handle on-call responsibilities

All features of Grafana OnCall are included with an IRM user license. A Cloud Free subscription includes access for three users. Both the Pro and Advanced accounts include access to 5 users; additional users are billed at $20 per month at the time of writing.

In the next section, we will look at alert groups and how to set up integrations for inbound and outbound data flow. We will also explore the templating language used in Grafana OnCall and how to manage escalation chains.

Alert groups

All alerts that flow into Grafana OnCall are grouped into **alert groups**. These groups can consist of one or more individual alerts, and the grouping behavior is managed by the integration template that is being applied. We will discuss templating after looking at integrations. An alert group can be in one of the following states at any time – **firing**, **acknowledged**, **silenced**, or **resolved.** Actions taken by on-call engineers or escalation chains can transition the state of an alert group. An alert group will look like this:

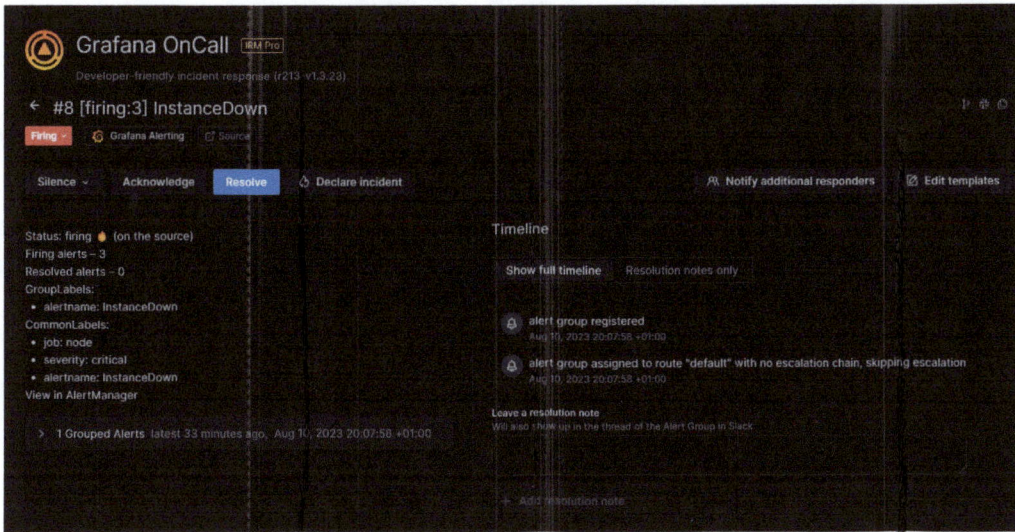

Figure 9.4 – The alert group anatomy

This web view mirrors the functionality available to on-call engineers via integrated communications channels, the mobile app, phone calls, and SMS. Alert groups can easily be acknowledged, unacknowledged, silenced, or resolved. Engineers can also notify additional responders if they need to bring in another team, declare an incident to trigger the processes in Grafana Incident (which is discussed later in this chapter), or combine alert groups if they are all related, meaning that cleaning up after an incident is much easier.

Inbound integrations

The tools to set up an **inbound integration**, or **alert source**, are under the **Integrations** option. These interactions are used to send alert information from an external source to Grafana OnCall. There are currently over 20 integrations, and inbound webhooks can be used to integrate with any system that can send them. Clicking on **New Integration** will begin the process of connecting to a new alert source; we will use Grafana Alerting as our source, as we have just explored it. First, give the integration a name and description, and then select an alert manager and a contact point. Grafana OnCall is able to integrate with any Prometheus-compatible alert manager; a default alert manager is configured in Grafana Cloud, called *Grafana*. Finally, click **Create Integration** and you will see the following screen:

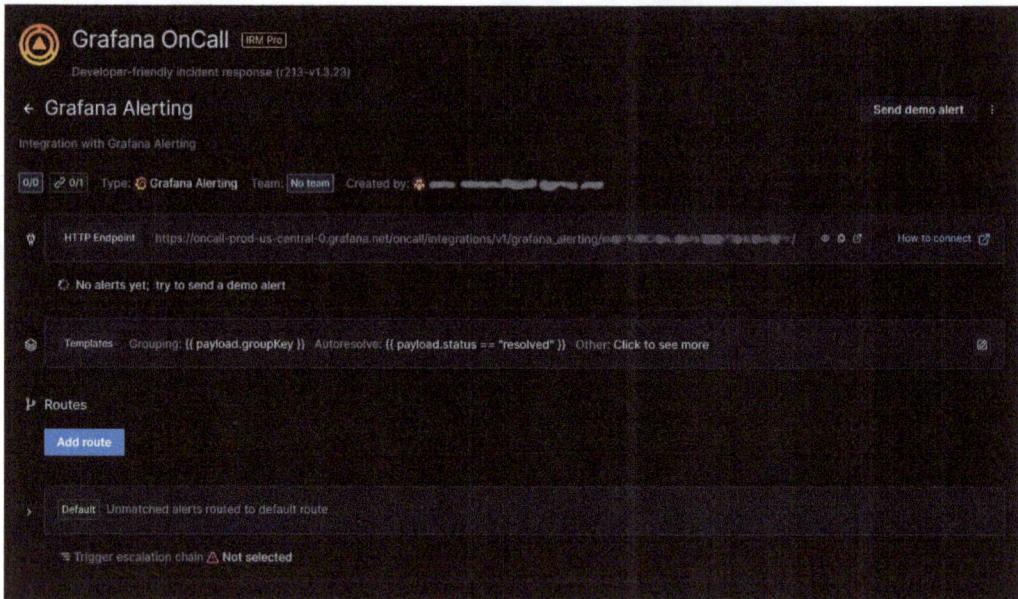

Figure 9.5 – Grafana alerting integration

The **HTTP endpoint** is used to configure the alert source to send alerts to. If you need assistance in configuring a specific integration, the **How to connect** link offers additional information. To test the integration, the **Send demo** alert will create a test alert.

The **Templates** block on the screen shows how OnCall will take a JSON payload and parse it. This uses Jinja2 templating to set various fields of the OnCall alert group. Each integration will have a different template to parse the unique payload sent by each source. In the Grafana Alerting template, we create a `grouping ID` field from the value in the `payload.groupKey` field. Similarly, the alert group will be resolved if `payload.status` is `resolved`. This means that the source of the alert can also send a resolution update as well.

The next sections, **Web**, **Phone**, **Slack**, **Telegram**, **Email**, and **MS Teams**, hold the template for how a notification about an alert group will be sent to these ChatOps integrations.

Adding routes allows you to select an escalation chain, based on a Jinja routing template for each alert group that originates from the integration. This is achieved by clicking on the **Add route** button. Routes also include the option to publish to any ChatOps integrations that have been configured.

Another important function available on the triple dot menu on the integration page is the ability to start a maintenance period. Maintenance can be either a debug, which silences all escalation, or standard maintenance, which collects all alerts into one alert group.

Let's cover templating and escalation chains next.

Templating

Jinja is a templating language with many useful features to parse multiple alerts in an alert group, enabling valuable information to be seen quickly in the messages sent to those on call. Here are some of the features and syntax of the language:

- **Loops**: The syntax for a loop is this:

```
{% for item in seq -%}
Do something with item from seq
{% else %}
Do something else if there are no items in seq
{%- endfor %}
```

Let's use an example. We will assume our payload is as follows:

```
{"results": [{"metric": "bigbadwolf", "value": 1},{"metric":
"littlepiggies", "value": 3},{"metric": "houses", "value": 1}]}
```

We can then use the following Jinja template:

```
*Values:*
{% for item in results-%}
{{ item['metric'] }}: '{{ item['value'] -}}'{{ "\n" }}
{%- endfor %}
```

This will result in this output:

```
*Values:*
bigbadwolf: '1'
littlepiggies: '3'
houses: '1'
```

- **Conditions**: Conditions can be constructed with this syntax:

```
{% if field == condition1 %}
Do something
{% elif field == condition2 %}
Do something different
{% else %}
Do something else if condition 1 and 2 didn't match
{% endif %}
```

- **Functions**: Jinja offers a comprehensive list of built-in functions that can be used in templates. Grafana has also added a few additional functions:

  - `time`: The current time

  - `tojson_pretty`: JSON prettified

- iso8601_to_time: Converts time from iso8601 (2015-02-17T18:30:20.000Z) to datetime format

- datetimeformat: Converts time from datetime to the given format (%H:%M/%d-%m-%Y by default)

- regex_replace: Performs a regex find and replace

- regex_match: Performs a regex match, and returns True or False

- b64decode: Performs a Base64 string decode

- **White space management**: Jinja templates blocks of text, so it's sometimes useful to add white space in a template for readability but have it stripped in the formatted message. Jinja offers the option to add a minus sign (-) to the start or end of a block to remove all white space before or after it. If seq = [1,2,3,4,5,6,7,8,9], we can write this as follows:

```
{% for item in seq -%}
    {{ item }}
{%- endfor %}
```

This will strip the white space before and after the item, and this is rendered as follows:

```
123456789
```

Without this white space management, it would be rendered as follows:

```
1
2 ...
```

Jinja also offers the ability to trim functions, which can remove white space as well. If you want to maintain white space, adding a plus sign (+) will indicate that it should be retained.

For more information on Jinja, please check out the website: https://jinja.palletsprojects.com.

Now that we understand how to use templates to format payloads and messages, let's take a look at escalation chains.

Escalation chains

Escalation chains let us set up standard workflows for alert groups. This is great for routing alerts based on the severity or content of the alert, the time of day, or other factors. There are a number of steps that can be set up:

- Wait: Wait for a specified amount of time and then continue to the next step.

- Notify users: Send a notification to a user or a group of users.

- `Notify users from on-call schedule`: Send a notification to a user or a group of users from an on-call schedule.

- `Resolve incident automatically`: Resolve the alert group right now with the status `Resolved automatically`.

- `Notify whole Slack channel`: Send a notification to a Slack channel.

- `Notify Slack User Group`: Send a notification to a Slack user group.

- `Trigger outgoing webhook`: Trigger an outgoing webhook.

- `Notify users one by one (round robin)`: Each notification will be sent to a group of users one by one, in sequential order and round-robin fashion.

- `Continue escalation if current time is in range`: Continue the escalation only if the current time is in a specified range. This can be used to pause escalations outside of working hours.

- `Continue escalation if >X alerts per Y minutes` (beta): Continue the escalation only if it passes some threshold.

- `Repeat escalation from beginning` (5 times max): Loop the escalation chain.

When a notification is sent to a user, either directly, via an on-call schedule, or via a round-robin, then the user's personal notification steps are followed. These can be managed by the user. The user's page will highlight the status of notification steps for all users; any users who have not configured notifications will be marked with a warning.

Outbound integrations

Grafana OnCall offers several ways to perform **outbound integration**, which involves integrating external tools so that outbound messages can be sent, either to a messaging tool or any system that can receive a webhook. There are two types of such integration:

- **ChatOps,** which are integrations that include Slack, Telegram, and MS Teams. These are configured via **Settings | ChatOps.**

- **Webhooks**: These outgoing Webhooks provide the ability to integrate with any system that can receive them, and they are triggered from events in OnCall.

The following screenshot shows how to set up a webhook in Grafana OnCall:

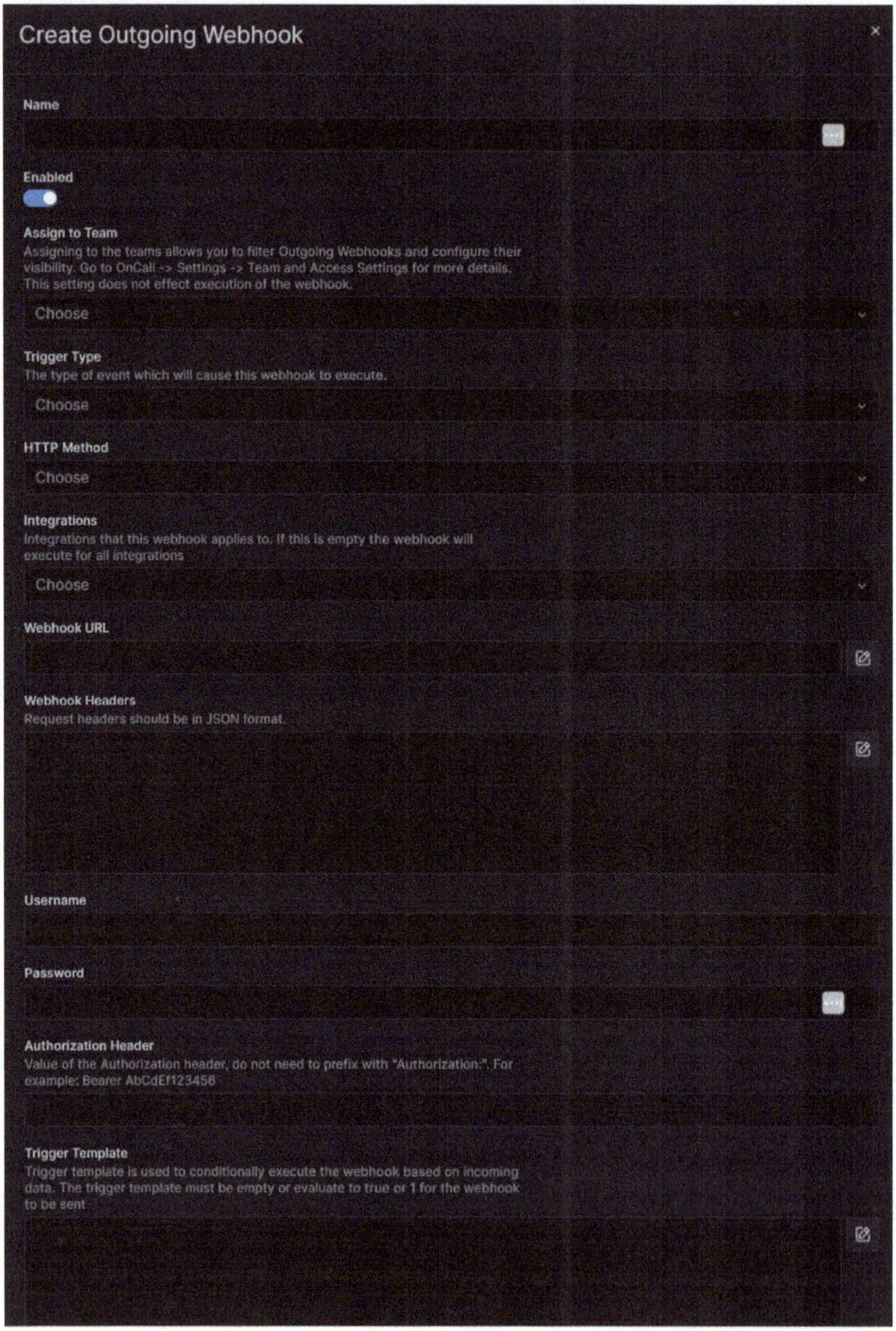

Figure 9.6 – Configuring an outbound webhook

Webhooks can be triggered from an escalation step, or when an alert group is created or transitions to particular states.

Schedules

Schedules are the way Grafana OnCall manages who is on call from each team. These are very easy to set up, and they offer you the ability to have a standard rotation schedule and set up any overrides. Schedules will also notify a Slack channel about the current on-call shift and any unassigned shifts. The **New Schedule** screen is shown in the following screenshot:

Figure 9.7 – Setting up a new schedule

Now, let's look at Grafana Incident.

Grafana Incident

The final major component of Grafana's IRM offering is **Incident**. This tool helps simplify and automate many aspects of the incident management tactical plan. The tool integrates with an organization's chat tooling, such as Slack, and offers you the ability for team members to declare incidents in that tool.

Once an incident is declared, Grafana Incident can automatically start a video conference, update status page tools, add context to the incident timeline from GitHub, and so on, depending on the integrations that have been configured. Team members of the incident can specify who takes what role in the incident and specify and assign tasks during the incident. As the incident evolves, Grafana Incident will record important information in a timeline, which can be published and easily reviewed in regular post-incident reviews.

To begin using Grafana Incident in Grafana cloud, an administrator needs to agree to a few integrations being set up initially. It is also good to set up integrations with a messaging tool such as Slack and a video conferencing tool such as Zoom, as Grafana Incident will use these when an incident is declared. A new incident bridge will be created in video conferencing, and the tool can record chat messages when instructed so that they are available for a post-incident review. Where applicable, integrations with Statuspage, GitHub, and Jira should also be configured; these can update Statuspage, record the state of pull requests and issues, and manage bug tickets, respectively. We expect the list of available integrations to expand as this tool matures.

Let's look at how Grafana Incident is used during an incident:

- **During an incident**: During an incident, a commander and multiple investigators can be assigned. Predefined tags can be associated with the incident, and a severity can be set. Investigators can then record text updates in the incident timeline, as well as relevant queries, alerts, and panels visited. The tool will then collect all relevant chats, text updates, queries run, alerts that were fired, and panels that were used during the incident. Investigators can also fire outbound webhooks that have been configured. The incident screen includes a task list, and links to relevant resources can also be attached.

- **After an incident**: The information collected during an incident is collated into a timeline of the incident. The timeline for this incident can then be reviewed. Grafana Incident also collates standard metrics, which can be viewed at a higher level on the **Insights** page.

  The **Insights** page shows high-level metrics for all incidents in the last 90 days by default. It is best practice for the silver leadership teams in organizations to review this page as part of a regular formal process, reporting to gold teams. This helps to ensure that incidents are being handled well and remediation work is being scheduled by teams, reducing the frequency and impact of incidents on an organization.

Grafana offers several AI and **machine learning (ML)** tools to help incident management:

- **Suggestbot**: This tool uses **Natural Language Processing (NLP)** to analyze the conversation during an incident and will suggest dashboards that have titles that are related to the subject that is being discussed.

- **OpenAI integration**: This is a public preview tool at the time of writing. Its aim is to speed up the tedious process of writing post-incident summaries. The tool uses OpenAI's ChatGPT to distill the incident timeline into a summary, which can be fine-tuned. An OpenAI account is required to use this integration. This integration can produce the following:

  - A summarized description of incidents

  - An event timeline of what happened leading up to an incident

  - Details of the actions that were taken to resolve an incident

- **ML Forecasting**: This tool will forecast the future state of metrics based on the learned models of past states. This is only available for metric data (either from Prometheus, Graphite, or Loki metrics queries). These forecasts can be used in dashboards or to drive alerts.

- **ML Outlier Detection**: This tool builds a model of what *normal* looks like for a metric and will highlight when metrics are outside of this normal range. These outliers can be used in alerting.

- **ML Sift**: Sift is also a public preview tool. This is a powerful tool for incident management, as it will interrogate telemetry from infrastructure and help identify critical details that may be drowned out in the noise of an incident. Sift will look for the following:

  - Patterns in error logs

  - Crashes in Kubernetes clusters

  - Noisy Kubernetes nodes, which have high loads

  - Containers that have resource throttling

  - Deployments that occurred recently

  - Slow requests seen in Tempo

You should now feel confident in being prepared for and responding to incidents using the tools provided by Grafana.

Summary

In this chapter, we have seen how to establish a great incident management process, which will help you evaluate and work on your organization's own process. We have also explored SLIs, SLOs, and SLAs and how to use them, seeing immediately whether a service is responding successfully or not. You have gained the skills to select appropriate SLIs, allowing you to transparently share with the rest of your organization whether the services you are responsible for are behaving as expected. In turn, this transparency helps the organization identify quickly where problems are and target resources to address them.

Finally, we looked at the tools offered by Grafana for incident management, seeing how to configure and use them to support great incident management processes.

The next chapter will look at how we can use the tools provided by Grafana and OpenTelemetry to automate the processes of collecting, storing, and visualizing data in an observability platform.

10

Automation with Infrastructure as Code

This chapter will explore how to use **Infrastructure as Code (IaC)** tools to automate the management of the various components of the Grafana observability platform. We will focus on **Ansible**, **Terraform**, and **Helm**, which allow teams to manage many aspects of their systems repeatably and automatically. This chapter divides the platform into the *collection and processing* layer, the *storage* layer, and the *visualization* layer and will outline how to automate each of these components. This chapter will provide the technical tools to create an easy-to-manage and very scalable observability platform, and combined with the information in *Chapter 11*, you will be well placed to lead your organization in easily leveraging the power of observability.

In this chapter, we're going to cover the following main topics:

- Benefits of automating Grafana
- Introducing the components of observability systems
- Automating collection infrastructure with Helm or Ansible
- Getting to grips with the Grafana API
- Managing dashboards and alerts with Terraform or Ansible

Technical requirements

This chapter involves working with Ansible, Terraform, and Helm, and it is recommended that you install them before you start reading. The chapter will also discuss a couple of concepts that you should have at least a passing familiarity with:

- Kubernetes objects
- The Kubernetes operator pattern

Benefits of automating Grafana

Observability tooling combines the collection, storage, and visualization of telemetry from many applications, infrastructure services, and other components of systems. Automation offers us a way of providing a testable, repeatable way of delivering these needs. Using industry-standard tools such as Helm, Ansible, and Terraform helps us maintain these systems in the long term. There are a lot of benefits to using automation, including the following:

- It reduces the risks associated with manual processes.

- Domain experts can provide automation for systems that developers interact with. This gives development teams confidence that they are using unfamiliar systems correctly. This knowledge comes in the following forms:

  - Data architecture for telemetry

  - Repeatable system architecture

  - Best practices for managing data visualizations

- By providing automation, domain experts are able to focus on higher-value work by letting teams self-serve using more straightforward and user-friendly systems.

- It provides a golden path so development teams can adopt observability easily and quickly and spend more time focusing on value-adding activities.

- It allows for easy scaling of best practices and operational processes. This is especially important for organizations that are growing, where a process that may work with a handful of teams does not scale to dozens of teams.

- It ensures that cost information is always attributable.

Now that we've introduced why you would want to use automation, let's have a look at the components that make up an observability platform so we can easily automate the different aspects of the system.

Introducing the components of observability systems

Observability systems consist of many components involved in producing, consuming, transforming, storing, and using data. Over the course of this chapter, we will split these components into four distinct systems to be clear about which aspect of observability platforms we are discussing. The different aspects of automation will be of interest to different audiences. The systems we will discuss are as follows:

1. **Data production systems**: These are the systems that generate data. The applications, infrastructure, and even components of the data collection system will produce data. Let's look at the key features:

   - These systems are managed by developers such as *Diego*, or by operations experts such as *Ophelia* (refer to *Chapter 1* for details on these personas).

- These systems are tested as part of the application- or component-testing process. If a data schema is in use, this can be validated using a tool such as JSON Schema.

2. **Data collection systems**: These systems collect the logs, metrics, and traces generated by data-producing systems. They typically offer tools for transforming data. Their key features are the following:

   - These systems are often run by specialist operations teams, observability engineers, site reliability engineers, or platform engineers

   - These systems are provisioned infrastructure

   - Automation involves the use of IaC tools and static analysis tools (where available) to validate the infrastructure

3. **Data storage systems**: These are the systems that store data and make it searchable. If your observability platform leverages SaaS tools, these systems will be provided by your vendor. Loki, Prometheus, Mimir, and Tempo are all examples of storage systems. Some of the important features of these systems include the following:

   - These systems are often managed by dedicated third parties, but when they are managed within an organization, they will typically be managed by the same team as the data collection system

   - These systems are provisioned infrastructure

   - Automation involves the use of IaC to provision on-prem resources, or leverages SaaS tooling such as Grafana Labs with IaC configuration

4. **Data visualization systems**: These are the systems that allow users to search the data stored in the storage systems and produce visualizations, alerts, and other methods of understanding the data. Grafana is an example of a visualization system. The following are some important features of such systems:

   - The management of this layer is typically a shared responsibility. The developers and operators who manage a particular system should be empowered to take ownership of their dashboards. The team managing the collection and storage layers will typically be the team empowering the rest of the organization.

   - These systems are provisioned infrastructure.

   - Automation involves the use of IaC to provision on-prem resources, or leveraging SaaS tooling such as Grafana Labs, with IaC configuration.

In this chapter, we will discuss systems 2, 3, and 4 in the preceding list. *System 1*, while important, is a very broad area and the automation strategies differ for different types of data producers. However, in most cases, teams can rely on the testing done by the libraries they consume, or the third-party systems they run.

Let's start by looking at how we can use Terraform or Ansible to deploy data collection systems.

Automating collection infrastructure with Helm or Ansible

Automating the installation of the infrastructure used to collect telemetry is a critical piece of building a great observability platform. The tools to support this depend on the infrastructure you are deploying to. In this section, we will examine the installation of the **OpenTelemetry Collector** and **Grafana Agent** using the following tools:

- **Helm** is a tool for packaging and managing Kubernetes applications. A Helm chart contains all the configuration files for the various Kubernetes components needed for an application, and typically handles setting the variables for the application. We will be using Helm in a Kubernetes environment.

- **Ansible** is a tool for standardizing operations into repeatable playbooks. It uses simple YAML configuration files to define the actions to be taken and leverages OpenSSH to connect to the target servers on which the actions are to be taken. We'll be using Ansible in a virtual or bare-metal environment, but it can be used to manage Kubernetes environments as well.

> **Important note**
>
> OpenTelemetry and Grafana both offer a Kubernetes operator, which can be installed using Helm. We will provide an overview of these tools as well.

Now let's look at how we can use Helm and Ansible to automate the installation of the OpenTelemetry Collector and Grafana Agent.

Automating the installation of the OpenTelemetry Collector

In this book, we have been using the OpenTelemetry Collector to collect data from the OpenTelemetry demo application and send it into our Grafana instance. First, we'll use the configuration we have already deployed to explore using the Helm chart made by OpenTelemetry to deploy the collector into a Kubernetes cluster.

OpenTelemetry Collector Helm chart

We first installed the OpenTelemetry Helm chart in *Chapter 3*, and then updated the configuration in *Chapters 4*, *5*, and *6*. OpenTelemetry provides detailed information about the configuration options that are available in its Git repository at `https://github.com/open-telemetry/opentelemetry-helm-charts/tree/main/charts/opentelemetry-collector`.

Let's look at the final configuration that we applied in *Chapter 6* and see how we configure the OTEL Helm chart. You can find this file in the Git repository at `/chapter6/OTEL-Collector.yaml`.

The first configuration block we'll look at is `mode`, which describes how the collector is going to be deployed in Kubernetes. The options available are `deployment`, `daemonset`, and `statefulset`. Here, we use the `deployment` option:

```
mode: deployment
```

Let's explore the available options in some detail:

- A `deployment` is deployed with a fixed number of Pods, which is the `replicaCount` in Kubernetes terms. For our reference system, we used this mode as we know the system will be deployed to a single-node Kubernetes cluster, and it allows us to combine presets that would usually be used independently in a multi-node cluster.

- A `daemonset` is deployed with a collector to every node in a cluster.

- A `statefulset` is deployed with unique network interfaces and consistent deployments.

We discuss selecting the appropriate mode in *Chapter 11* when we discuss architecture. These deployment modes can also be combined to provide specific functionality in a Kubernetes cluster.

The next configuration block we'll look at is `presets`:

```
presets:
  logsCollection:
    enabled: true
    includeCollectorLogs: false
  kubernetesAttributes:
    enabled: true
  kubernetesEvents:
    enabled: true
  clusterMetrics:
    enabled: true
  kubeletMetrics:
    enabled: true
  hostMetrics:
    enabled: true
```

As you can see, this configuration involves simply enabling or disabling different functions. Let's look at the parameters in detail:

- The `logsCollection` parameter tells the collector to collect logs from the standard output of Kubernetes containers. We are not including the collector logs as this can cause a logging cascade, where the collector reads its own log output and writes the collected logs to that output, which are then read again. In real-life setups, it is recommended to only use the `logsCollection` parameter in `daemonset` mode.

- The kubernetesAttributes parameter collects Kubernetes metadata as the collector receives logs, metrics, and traces. This includes information such as k8s.pod.name, k8s.namespace.name, and k8s.node.name. The attribute collector is safe to use in all modes.

- The kubernetesEvents parameter collects the events that occur in the cluster and publishes them in the log pipeline. Effectively, every event that occurs in the cluster receives a log entry in Loki with this configuration. Cluster events include things such as Pod creations and deletions, among others. It's best practice to use kubernetesEvents in the deployment or statefulset modes to prevent the duplication of events.

- The three metrics options collect metrics about the system:

  - clusterMetrics looks at the full cluster. This should be used in deployment or statefulset modes.

  - kubeletMetrics collects metrics from the kubelet about the node, Pods, and containers it is managing. This should be used in daemonset mode.

  - hostMetrics collects data directly from the host, such as CPU, memory, and disk usage. This should be used in daemonset mode.

We'll skip over a few blocks that are standard Kubernetes configurations and consider the config block next. The config block has a few subblocks:

- The telemetry pipeline includes the following:

  - receivers: Receivers are at the start of a pipeline. They receive data and translate it to add it to the pipeline for other components to use.

  - processors: Processors are used in a pipeline to carry out various functions. There are supported processors and contributed processors available.

  - exporters: Exporters come at the end of a pipeline. They receive data in the internal pipeline format and translate it to send the data onwards.

  - connectors: These combine receivers and exporters to link pipelines together. connectors act as exporters to send the data from one pipeline onwards, and as receivers to take that data and add it to another pipeline.

- Separate to the pipeline are extensions, which add additional functionality to the collector, but do not need access to the telemetry data in the pipelines.

- Finally, there is a service block, which is used to define the pipelines and extensions in use.

The only extension we are using in our config block is the health_check extension:

```
config:
  extensions:
    health_check:
```

```
        check_collector_pipeline:
          enabled: false
```

This enables an endpoint that can be used for a liveness and/or readiness probe in the Kubernetes cluster. This is helpful for you to be able to see easily whether the collector is working as expected.

In our `receivers` block we have configured two receivers, `otlp` and `prometheus`:

```
config:
  receivers:
    otlp:
      protocols:
        http:
          endpoint: 127.0.0.1:4318
          cors:
            allowed_origins:
              - "http://*"
              - "https://*"
    prometheus:
      config:
        scrape_configs:
          - job_name: 'opentelemetry-collector'
            tls_config:
              insecure_skip_verify: true
            scrape_interval: 10s
            scrape_timeout: 2s
            kubernetes_sd_configs:
              - role: pod
            relabel_configs:
              - source_labels: [__meta_kubernetes_pod_annotation_
prometheus_io_scrape]
                action: keep
                regex: "true"
              - source_labels: [__address__, __meta_kubernetes_pod_
annotation_prometheus_io_port]
                action: replace
                target_label: __address__
                regex: ([^:]+)(?::\d+)?;(\d+)
                replacement: $$1:$$2
```

Let's look at these receivers more closely:

- The OTLP receiver configures our collector instance to expose port 4318 on 127.0.0.1 on the Kubernetes node, which allows the demo applications to submit telemetry easily.

- The Prometheus receiver is used to collect metrics from the collector itself. This receiver config shows an example of relabeling, where we take `meta_kuberentes_pod_annotation_prometheus_io_port` and rename it with `__address__`, which is the standard used in OTLP.

In our configuration, we have set up the `k8sattributes`, `resource`, and `attributes` processors. The `k8sattributes` processor extracts attributes from the kubelet and adds them to the telemetry in the pipelines. The `resource` and `attributes` processors will insert or modify the resource or attributes respectively. We'll not discuss these concepts in detail, but resources are used to identify the source that is producing telemetry.

In our configuration, we are using both the `spanmetrics` and `servicegraph` connectors:

```
config:
  connectors:
    spanmetrics:
      dimensions:
      - name: http.method
        default: GET
      - name: http.status_code
      namespace: traces.spanmetrics
    servicegraph:
      latency_histogram_buckets: [1,2,3,4,5]
```

Both `connectors` are used to export the data from the `traces` pipeline and receive it in the `metrics` pipeline. `spanmetrics` collects the **request, error, and duration (RED)** metrics from the span data (we introduced RED in *Chapter 9*). `servicegraph` generates metrics that describe the relationship between services, these metrics allow the service graphs to be shown in Tempo.

The final subblock in our `config` block is `services`. This subblock defines the extensions to be loaded and the configuration of the pipelines. Each pipeline (`logs`, `metrics`, and `traces`) defines the `receivers`, `processors`, and `exporters` used. Let's look at the `metrics` pipeline as it is the most complex:

```
metrics:
      receivers:
        - otlp
        - spanmetrics
        - servicegraph
      processors:
        - memory_limiter
        - filter/ottl
        - transform
        - batch
      exporters:
```

```
      - prometheusremotewrite
      - logging
```

The receivers are OTLP, spanmetrics, and servicegraph as discussed previously. We then instruct the pipeline to use the memory_limiter, filter, transform, and batch processors, in the order listed. You may notice that our filter is named ottl using the syntax of processor/ name, which is useful when you need to use the same processor with different configurations. Finally, the pipeline uses the prometheusremotewrite exporters and logging to output data.

You may have noticed that the exporters are not defined in the /chapter6/OTEL-Collector. yaml file – this is because they are defined in /OTEL-Creds.yaml, and this highlights a very useful feature of Helm, which is the ability to separate out configuration files based on their function. When we install the Helm chart, we use a command such as the following:

```
helm install owg open-telemetry/opentelemetry-collector --values
chapter3/OTEL-Collector.yaml --values OTEL-Creds.yaml
```

The -f or --values option can be used multiple times for multiple YAML files – if there are conflicts, then precedence is always given to the last file used. By structuring the YAML files in such a way, we can split the full configuration in ways that allow us to protect secret information, such as API keys, while still making our main configuration easily available. We can also use this feature for other purposes such as overriding a default configuration in a test environment. It's important to be careful with precedence here as duplicate arrays will not be merged. Deploying the collector in this way is fantastic in a lot of situations. However, it has a limitation – any time there is a change to the configuration, or a new version of the collector is to be installed, a Helm install or upgrade operation needs to be carried out. This introduces the need for a system that has knowledge of and access to each cluster in which the collector will be deployed, which can introduce bottlenecks and security risks. Let's look at the OpenTelemetry Operator, which offers solutions to these problems.

OpenTelemetry Kubernetes Operator

OpenTelemetry also offers a Kubernetes operator to manage both the OpenTelemetry Collector and allow for auto-instrumentation of workloads. This operator is still in active development and the feature set is expected to increase.

Kubernetes operators use **Custom Resource Definitions** (**CRDs**) to provide extensions to the Kubernetes API, which are used by the operators to manage the system. The advantages of using an operator include the following:

- The complex logic involved in managing the OpenTelemetry Operator or the auto-instrumentation system can be designed by the experts working on the OpenTelemetry projects. For the person responsible for managing an OpenTelemetry installation, the operator offers a defined CRD specification against which to validate the proposed configuration in a CI/CD pipeline.

- The OpenTelemetry operator also allows for limited automated upgrades. Minor and major version updates still need to be applied via a Helm upgrade due to the possibility of breaking changes.

- It can be combined with GitOps tooling to move from a solution where a central system must know each cluster and have the necessary credentials to deploy to them, to a solution where each cluster reads the desired configuration from a central version-controlled repository.

- The operator really excels when it comes to making the OpenTelemetry auto-instrumentation easily accessible to applications by adding annotations. For the majority of use cases, auto-instrumentation will provide ample metric and trace telemetry to understand an application.

The OpenTelemetry Collector can also be installed on virtual or bare-metal servers, this process can be automated with tools such as Ansible. Let's see how you might approach this.

OpenTelemetry and Ansible

OpenTelemetry does not provide an official collection for Ansible. It provides packaged versions of the collector for Alpine-, Debian-, and Red Hat-based systems as `.apk`, `.deb`, and `.rpm` files respectively. Using `community.general.apk`, `ansible.builtin.apt`, or `ansible.builtin.yum`, the package can be installed with a configuration similar to the following:

```
- name: Install OpenTelemetry Collector
  ansible.builtin.apt:
    deb: https://github.com/open-telemetry/opentelemetry-collector-
releases/releases/download/v0.85.0/otelcol_0.85.0_linux_amd64.deb
```

With the package installed, the only other thing to do to configure the collector is to apply a configuration file. The default configuration file is located at `/etc/otelcol/config.yaml` and is used when systemd starts the collector. Ansible can overwrite this file or modify it in place. This could be done with the following configuration:

```
- name: Copy OpenTelemetry Collector Configuration
  ansible.builtin.copy:
    src: /srv/myfiles/OTEL-Collector.yaml
    dest: /etc/otelcol/config.yaml
    owner: root
    group: root
    mode: '0644'
```

We have looked at the OpenTelemetry agent a lot during this book. One major reason for this is that OpenTelemetry also offers the Demo application we have used to produce realistic sample data. Next, let's take a look at Grafana Agent.

Automating the installation of Grafana Agent

Grafana produces its own agent, which is recommended by Grafana for use with its cloud platform. Grafana Agent also provides automation options, which we will introduce in this section.

Grafana Agent Helm charts

Grafana offers two Helm charts, the **grafana-agent chart** and the **agent-operator chart**. Similar to the OpenTelemetry charts, the Agent chart allows for a direct installation of the Grafana Agent on a Kubernetes cluster. The Operator chart deploys CRDs into the cluster and then manages the state of the Agent based on these definitions. At the time of writing, the Grafana Agent Operator offers CRDs for metrics and log collection. For full details of the `grafana-agent` chart and the configuration options, please check out the Helm chart documentation at `https://github.com/grafana/agent/tree/main/operations/helm/charts/grafana-agent`. Similarly, the documentation for the `agent-operator` chart can be found at `https://github.com/grafana/helm-charts/tree/main/charts/agent-operator#upgrading-an-existing-release-to-a-new-major-version`. Details of the available CRDs are documented in the Operator architecture documentation at `https://github.com/grafana/agent/blob/v0.36.2/docs/sources/operator/architecture.md`.

Grafana Agent and Ansible

Unlike OpenTelemetry, Grafana maintains an Ansible collection (`https://docs.ansible.com/ansible/latest/collections/grafana/grafana`) that includes tools to manage collection, storage, and visualization systems, and we will revisit it in later sections.

The included `grafana_agent` role is used to manage data collection and will install the agent on Red Hat, Ubuntu, Debian, CentOS, and Fedora distributions. This role can be used as follows:

```
- name: Install Grafana Agent
  ansible.builtin.include_role:
    name: grafana_agent
  vars:
    grafana_agent_logs_config:
      <CONFIG>
    grafana_agent_metrics_config:
      <CONFIG>
    grafana_agent_traces_config:
      <CONFIG>
```

The configuration for logs, metrics, and traces is specific to that telemetry type and the documentation available from Grafana covers using that configuration to manage the Agent.

Getting to grips with the Grafana API

Grafana offers a full-featured API for both Grafana Cloud and Grafana itself. This API is the same API used by the frontend, which means that we can also drive the functions of Grafana using either direct API calls in a script, or an IaC tool such as **Terraform**. We'll start by having a high-level look at the APIs available in Grafana Cloud and Grafana, then we'll look at the Grafana Terraform module and the Ansible collection, and see how to use them to manage a Grafana Cloud instance.

Exploring the Grafana Cloud API

The Grafana Cloud API is used to manage all aspects of a Grafana Cloud SaaS installation. Let's have a high-level look at the functions provided by the Grafana Cloud API:

- **Access policies and tokens**: These API endpoints manage authentication and authorization resources. Here are their major functions:

  - Create, read, update, and delete functions for access policies

  - Create, read, update, and delete functions for tokens

  Tokens must be associated with an `accessPolicyId`, which is the unique ID for an access policy object.

- **Stacks**: These endpoints manage Grafana Cloud stacks. The following are their key functions:

  - Create, read, update, and delete functions for stacks

  - Restart Grafana on a specific stack

  - List data sources on a specific stack

- **Grafana plugins**: These endpoints manage plugins installed on Grafana instances related to a stack. Their functions include creating, reading, updating, and deleting functions for Grafana plugins installed on a specific stack.

- **Regions**: These API endpoints list the Grafana Cloud regions available. They are used to read functions for the available Grafana Cloud regions that can be used to host a stack.

- **API keys**: These endpoints were for managing Cloud API keys and their major functions are the create, read, and delete functions for API keys. These endpoints are now deprecated as Grafana has moved to authentication techniques using access policies and tokens. API key endpoints will be removed in a future update.

We will discuss access policies and tokens in more detail in *Chapter 11* where we discuss **access levels** as part of architecting a great observability platform. All the Grafana Cloud endpoints have an associated access policy, and the token used must be authorized with that policy for a successful response.

More detailed information is available in Grafana's documentation at `https://grafana.com/docs/grafana-cloud/developer-resources/api-reference/cloud-api/`, including information on the parameters and details needed in a request as well as example requests and responses.

We've now reviewed the API endpoints. Next, we'll discuss Grafana's offerings of both a Terraform provider and the Ansible collection, which can be used to interact with the aforementioned APIs using IaC automation.

Using Terraform and Ansible for Grafana Cloud

Grafana provides both a Terraform provider and an Ansible collection for use in managing organizations' Cloud instances. Let's explore how we can use these tools with the Grafana Cloud API to manage a Grafana Cloud instance.

The Grafana Terraform provider

The **Grafana Terraform provider** has resources that match the Cloud API endpoints we have discussed. The provider also offers resources for other Grafana API endpoints, which we will cover when we discuss managing dashboards and alerts later in this chapter. The official documentation for the provider can be found on the Terraform Registry at `https://registry.terraform.io/providers/grafana/grafana/latest/docs`.

> **Important note**
> This chapter was written with version 2.6.1 of the Grafana Terraform provider.

Here are some of the commonly used Terraform resources with examples of their use:

- First, let's have a look at using the provider. The `grafana_cloud_stack` data provider is used to find a stack called `acme-preprod`, which we will use later to specify where our access policy is to be created:

```
data "grafana_cloud_stack" "preprod" {
  slug = "acme-preprod"
}
```

- The `grafana_cloud_access_policy` resource allows us to create an access policy. Here, we set our `region` value to us, along with a name and a display name. Finally, we specify the scope of the policy – in this case, we want to be able to write logs, metrics, and traces. The `stack` ID we found earlier is then used to specify where to create this access policy:

```
resource "grafana_cloud_access_policy" "collector-write" {
  region       = "us"
  name         = "collector-write"
```

```
display_name = "Collector write policy"

scopes = ["logs:write", "metrics:write", "traces:write"]

realm {
  type       = "stack"
  identifier = data.grafana_cloud_stack.preprod.id
}
}
```

- Finally, the `grafana_cloud_access_policy_token` resource can be used to create a new token. We specify a region, the access policy to use, and a name. The token will then be able to be read from `grafana_cloud_access_policy_token.collector-token.token`:

```
resource "grafana_cloud_access_policy_token" "collector-token" {
  region            = "us"
  access_policy_id = grafana_cloud_access_policy.collector-
write.policy_id
  name              = "preprod-collector-write"
  display_name      = "Preprod Collector Token"
  expires_at        = "2023-01-01T00:00:00Z"
}
```

This isn't the only thing we can do with the Grafana Terraform provider: we'll consider another example later in this chapter when we examine managing dashboards and alerts. Typically, this would be combined with another provider to record this newly created token in a secrets management tool, such as **AWS Secrets Manager** or **HashiCorp Vault**, where it can be accessed whenever a collector is deployed.

Let's look managing a Grafana Cloud system with another IaC tool provided by Grafana, the **Grafana Ansible collection**.

The Grafana Ansible collection

The Grafana Ansible collection is not as feature rich as the Terraform provider for managing cloud instances. However, a lot of the functionality from the Grafana Cloud API can be accessed using the Ansible URI module. The official collection documentation is available on the Ansible site at `https://docs.ansible.com/ansible/latest/collections/grafana/grafana/`. A community-provided collection is also available, but will not be discussed here. The relevant documentation is available at `https://docs.ansible.com/ansible/latest/collections/community/grafana/index.html`.

> **Important note**
> This chapter was written with version 2.2.3 of the Grafana Ansible collection.

We'll look at managing a Grafana Cloud stack using Ansible. The `name` and `stack_slug` (this is the stack we are interacting with) values are set to the same string by convention. We then need to set the `region` value for the stack, along with the organization the stack will belong to using `org_slug`:

```
- name: Create preprod stack
  grafana.grafana.cloud_stack:
    name: acme-preprod
    stack_slug: acme-preprod
    cloud_api_key: "{{ grafana_cloud_api_key }}"
    region: us
    org_slug: acme
    state: present
```

We've so far looked at the API for Grafana Cloud. This API is great for managing an observability platform in Grafana Cloud, but a lot of teams will be more interested in managing dashboards, alerts, and other items in the Grafana UI. Grafana provides another API to manage objects in the Grafana UI – let's look at it now.

Exploring the Grafana API

While the Grafana Cloud API is only used to manage Grafana Cloud SaaS instances, the **Grafana API** is very far reaching as Grafana has a lot of functionality. These APIs can be used on both Grafana Cloud and locally installed Grafana instances.

Similar to the Grafana Cloud API, all endpoints use role-based access control. However, the Grafana API offers an additional authentication option: service accounts. Service accounts should be used for any application that needs to interact with Grafana.

As most teams will use a small subset of APIs frequently, we will only discuss a few APIs here. However, there are a lot of other APIs that can be used to automate the management of a Grafana instance. Let's take a closer look at some commonly used APIs:

- **Dashboard and Folder**: These endpoints manage dashboards and folders in Grafana. Their functions include the following:

  - Create, read, update, and delete functions for dashboards or folders

  - Create, read, update, and delete the tags on a dashboard

  Dashboards and folders have both an ID and a UID. The ID is only unique to a specific Grafana installation, while the UID is unique across installations.

- **Dashboard Permissions and Folder Permissions**: These APIs handle the access controls for dashboards and folders. They can be used to update permissions for a dashboard or folder. These endpoints use a UID, although there is a deprecated endpoint to update permissions for dashboards by ID. Permissions are set numerically: 1 = View, 2 = Edit, 4 = Admin. Permissions can be set for user roles or `teamId` values.

- **Folder/Dashboard Search**: This API allows users to search for dashboards and folders. This endpoint allows for complex searches using query parameters. The response is a list of matching objects including the UID of an object.

- **Teams**: These endpoints manage Teams in Grafana. They can be used to do the following:

  - Create, read, update, and delete teams

  - Get, add, and remove team members

  - Get and update team preferences

- **Alerting**: These complex APIs manage all of the aspects of alerts. This API manages everything to do with Grafana Alertmanager. It can be used to create, read, update, and delete alerts, alert rules, alert groups, silences, receivers, templates, and many more alerting objects.

These API endpoints are fantastic for managing Grafana. Grafana provides a detailed API reference at `https://grafana.com/docs/grafana/latest/developers/http_api/`.

Let's now look at how these API endpoints allow us to use the IaC tools of Terraform and Ansible to manage dashboards and alerts.

Managing dashboards and alerts with Terraform or Ansible

As dashboards are typically managed by the teams responsible for a service or application, it is best practice to separate the tooling to deploy dashboards from the tooling to manage observability infrastructure. We will discuss the practicalities of this in *Chapter 14*.

For managing dashboards, both Terraform and Ansible leverage the fact that Grafana dashboards are JSON objects, providing a mechanism to upload a JSON file with the dashboard configuration to the Grafana instance. Let's look at how this works.

The **Terraform** code looks like this:

```
resource "grafana_dashboard" "top_level " {
  config_json = file("top-level.json")
  overwrite = true
}
```

A collection of dashboard JSON files can be iterated over using the Terraform `fileset` function with a `for_each` command. This makes it very easy for a team to manage all its dashboards in an automated manner by saving the correct dashboard to the relevant folder.

The **Ansible** collection works in a very similar fashion:

```
- name: Create Top Level Dashboard
  grafana.grafana.dashboard:
    dashboard: "{{ lookup('ansible.builtin.file', ' top-level.json')
}}"
    grafana_url: "{{ grafana_url }}"
    grafana_api_key: "{{ grafana_api_key }}"
    state: present
```

Like the Terraform code, this could be iterated using the built-in `with_fileglob` function, allowing teams to manage all their dashboards in an automated fashion.

Unfortunately, with the latest changes to alerting in Grafana, the Ansible collection has not been updated to allow for alert management. With Terraform, you can manage Grafana Alerts in a very similar way to dashboards. Consider the following example code block:

```
resource "grafana_rule_group" "ateam_alert_rule" {
  name             = "A Team Alert Rules"
  folder_uid       = grafana_folder.rule_folder.uid
  interval_seconds = 240
  org_id           = 1
  rule {
    name           = "Alert Rule 1"
  }
  rule {
    name           = "Alert Rule 2"
  }
```

We've not included the full details of the two rules shown here as the required configuration block is too large. The Terraform documentation has a very clear example of a full alert at `https://registry.terraform.io/providers/grafana/grafana/latest/docs/resources/rule_group`.

Similar to the `grafana_dashboard` resource, `grafana_rule_group` can be iterated over by using a `dynamic` block to populate each rule from another source, such as a JSON file, for example. This makes the management of these rules significantly more user-friendly.

Summary

In this chapter, you were introduced to the benefits of automating the management of your observability platform, and saw how investing in good automation can allow subject-matter experts to shift repetitive and low-value work to others in the organization. We discussed the different aspects of observability platforms, being data production, collection, storage, and visualization. You also learned who is typically responsible for each aspect of the platform.

With the theory largely covered, we then went on to discuss how to manage the data collection layer, presenting an in-depth analysis of the OpenTelemetry Collector Helm configuration that has been used to collect data throughout this book. We contrasted the way Helm works with Ansible to deploy to a virtual or physical setup, and you gained valuable skills in understanding the structure of the management files used by each tool. We rounded out the automation of data collection systems by introducing the Helm chart and the Ansible collection for Grafana Agent. While we did not go into this in the same depth as the OpenTelemtry configuration, the skills required for managing the Grafana Agent are identical.

Our next topic was the Grafana API, where you learned that there are two APIs, one to manage the SaaS Grafana Cloud solution, and one to manage Grafana instances (both cloud and local). You were then introduced to the Terraform provider for Grafana and learned through specific examples how to manage both their cloud stacks and their Grafana instances. We then also looked at the Grafana Ansible collection and saw how it can be used to manage cloud stacks and Grafana instances as well as the data collection layer.

In the next chapter of this book, we will discuss how to architect a full observability platform that scales to the needs of your organization.

Architecting an Observability Platform

This chapter covers several topics related to **architecting** a great observability platform for teams in an organization to use. We will discuss how to structure data into **domains** to help find relevant data quickly in even the largest organizations, and how that relates to other aspects of the business, such as financial reporting and **business intelligence** (**BI**). Then, we will discuss architecting the four main system components of an observability platform: **data production**, **data collection**, **data storage**, and data uses such as **visualization** and **alerting**. We will cover how to link the architecture with the IaC tools that were discussed in *Chapter 10*. After that, we will discuss how to use various easily available tools to validate a design with local testing. These tools can also be used in CI/CD pipelines to validate the platform after a change has been implemented. We will discuss the **role-based access controls** (**RBACs**) that are implemented in Grafana and how to set them up to provide least-privilege access. Finally, we will briefly discuss how to architect connections with other systems that make use of the same telemetry, such as **security information and event management** (**SIEM**) or BI systems. This chapter is aimed at a senior technical audience who has experience in architecting platforms and systems.

In this chapter, we're going to cover the following main topics:

- Architecting your observability platform
- Proving theoretical designs (proof of concept)
- Setting the right access levels
- Sending telemetry to other consumers

Architecting your observability platform

Understanding and articulating the problem(s) your organization is trying to solve is the most critical and undervalued aspect of a well-architected observability platform. There are some common problems that organizations are trying to solve with observability, but every organization is different, and working with people such as *Masha* (senior leadership, as introduced in *Chapter 1*) to understand the business needs is a step that is often missed and can lead to complex problems in the future.

Here are some common problems that organizations face that can be solved with an observability platform:

- **Customer-affecting incidents**: These types of incidents could range from downtime to data breaches. These pose a compliance, operational, and reputational risk to the organization. The customers could be internal or external to the organization.

- **Understanding the organization's key performance indicators (KPIs)**: Organizations often want to have a clear understanding of the current state of their KPIs. These KPIs articulate whether the organization is doing well or whether something needs addressing.

- **Understanding how customers use products**: Understanding how customers interact with an organization's products can identify pain points and help guide a better experience. Offering great products gives the organization a competitive advantage.

- **Understanding the financial costs of serving customers**: This is commonly known as the **cost of goods sold (COGS)** and **operating expenses (OPEX)**.

In this section, we will consider how to architect the data structures used in an observability platform to support the organization's goals. We'll talk about the process of designing a system architecture and the considerations you should make to support the operational needs of the organization. Finally, we will think about designing management and automation processes so that following the best practices you establish becomes the easiest path for teams to take.

Defining a data architecture

A data architecture defines an organization's data assets and maps how data flows through the organization's systems. Most organizations will already have a data architecture in place, so it is worth discussing with the team responsible. In this section, we will discuss how the field names and data types in the observability platform need to match or be translatable into common fields across the organization.

Observability systems are inherently data systems. They collect, process, move, store, and use data. The data in an observability system is most valuable to the wider organization when it is compatible with other data systems so the organization can merge datasets. The crux of this is that when embarking on this journey, talk to people throughout the organization and find out who is responsible for the data architecture of the whole organization. If no one exists in that position, it can be raised with senior leadership as a hindrance to solving the problems they are trying to address. For example, when I was a junior engineer, I remember having many meetings discussing whether `tenantID` and `customerID` were different fields or not as there were two systems that used different names. Ultimately, it was decided they were different concepts so the business could capture the idea of principal and subsidiary organizations that were customers of the company. However, both systems then needed months of work to implement this wider concept. The logging platform also needed a lot of data model rebuilding to capture this new concept. This work would have been entirely avoidable by having someone responsible for the data model and defining the requirements early on.

It is common to use data models from other areas of the organization when implementing an observability platform. There is a step that should be completed by an architect where they translate external requirements into a requirements document detailing where fields should be recorded. For example, it may be a requirement of the financial data model to record cost centers. There are many ways to achieve this, such as the following:

- Requiring every log line, every metric, and every trace to include this information
- Requiring every service to be tagged with an `organization.costcenter` label
- Maintaining a lookup table of service name ↔ cost center

The requirements guide should be clear on how this will be achieved for teams who will be meeting the requirements. We recommend a document structure such as **MoSCoW**, which stands for **Must have, Should have, Could have, Won't have**.

Different telemetry types are best suited for different data types. Observability systems are also packed with features to gather data from other systems, such as Kubernetes object labels and cloud tags. These should form part of the data architecture. Here are some telemetry types and what they are best suited for:

- **Log fields in Loki**: Log fields are best suited to string data such as the following:

    - Application state fields in string format, such as *error* or *warn* states, for example, if an application queries data from another service and cannot connect
    - Organizational or business data fields, such as service name, customer ID, user ID, and so on
    - Low- to medium-cardinality indexed fields
    - High-cardinality unindexed fields
    - Link from application state to system state in traces

- **Metric fields in Prometheus or Mimir**: Metric fields are best suited to numeric data such as the following:

    - Application state fields in numeric format, such as the count of records processed since startup
    - Organizational or business data fields, including labels containing the service name, hostname, and so on
    - Low- to medium-cardinality fields, such as HTTP methods (GET, POST, PUT, etc.)

- **Trace fields in Tempo**: Trace fields are a complex data type that can handle data such as the following:

    - System state fields
    - High-cardinality fields

- Organizational or business data fields when added as an attribute, such as customer ID and user ID

- Cross-system fields

- Links to application state by using trace metrics

- **Kubernetes labels**: These are Kubernetes key-value pair data objects. They are used to record information such as the following:

  - Core organization fields, such as ownership and cost allocation

  - Linking the application to the infrastructure

  - These labels can be added to log, metric, and trace data as it is collected

- **Cloud vendor tags**: These are tags applied to infrastructure in a cloud vendor system. They can be used to record information such as the following:

  - Core organization fields, such as ownership and cost allocation

  - These labels can be added to log, metric, and trace data as it is collected

A lot of this data is standard across many organizations and industries, and the libraries that produce the data are well tested. There is one area of data production that is not tested by these tools though, and that is organization-specific fields. These are always organization-specific, but some common examples are user ID or customer ID. These fields, when used by the organization, can be very important, even being reviewed regularly by executive leaders. It is important that these are tested as bad data can lead to bad decisions. Any data architecture documents should highlight this need. There is a lot of technical detail in achieving this goal, which we will not go into in this book, but we would recommend this article from Martin Fowler, which gives clear instructions on producing organizational data in a testable way: `https://martinfowler.com/articles/domain-oriented-observability.html`.

We've now seen how to work within the organization to have a coherent data architecture that works with the infrastructure layer, the application layer, the observability layer, and the business layer. Let's now consider how to have a great system architecture for your organization's observability platform.

Establishing system architecture

In this section, we will consider the aspects of building a great observability system. We will see how to help software engineers with producing data. Then, we will consider how to collect that data, while providing engineers with a stable API. Finally, we'll discuss the storage and visualization of the data. The following list presents some questions to consider relating to these topics:

- How will data be produced?

  - What telemetry types (logs, metrics, traces, or others) are used?

- Will developers such as *Diego* be given standards for libraries?
- Should *system* or *state* data be separated from *business* data?

- How will data be collected?

  - What systems do you need to collect data from?
  - If a tool is changed, will every application need to be updated?
  - How much data will be collected?

- How will data be stored?

  - Will any local storage be provided? If so, how will the scale and cost of this be managed?
  - Is the storage managed per cluster or environment, or as a centralized system?
  - Will a third-party solution such as Grafana Cloud be used? If so, how is the cost allocated?

- How will visualizations be managed?

  - Will the system be fully open, so anyone can submit changes for any dashboard?
  - Will each team be responsible for their dashboards?
  - Will IaC tools be provided to help teams manage their dashboards?

- An additional question to consider for all of these is how the system employed handles failure

Let's cover these considerations in more detail, starting with architecting how applications produce data.

Data production

Data production details how applications and services produce data. Teams responsible for observability platforms should assist the teams who produce data in doing so with all the correct fields, practices, and standards. Common topics to cover are the following:

- Which telemetry types must be produced and which should or may be produced?
- Is organizational or business data being collected from the observability systems? If it is, what are the fields? Is a data domain used (e.g., `acme.cost_center` or `acme.department`)?
- Are developers expected to use libraries from a pre-approved list?
- What standards are used by applications to present data?

OpenTelemetry, while relatively young, is emerging as the standard in observability, with adoption across most vendors and systems. A suggested best practice for an application is to add instrumentation using the relevant OpenTelemetry SDK, as shown in the following diagram:

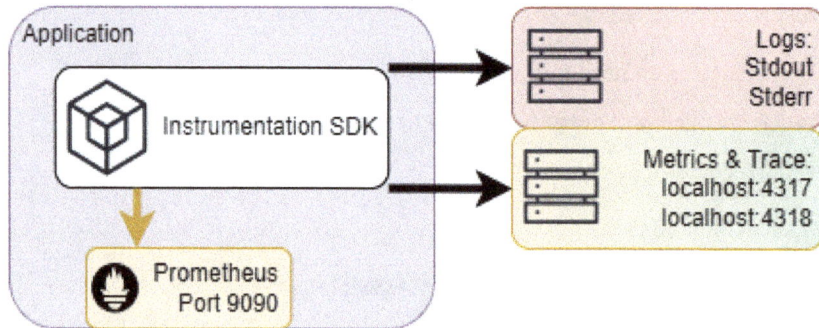

Figure 11.1 – Proposed application data production standard

Here, logs are produced on `stdout` and `stderr`. Metrics are published both to a Prometheus scrape endpoint and to an OpenTelemetry receiver via either gRPC on port `4317` or HTTP on port `4318`. Traces are also pushed to the OpenTelemetry receiver using the same ports.

Producing data is only part of the picture for a well-architected system. Next, let's look at how to design a system to collect all this data so it is useful to the organization.

Data collection

The data collection agents we have discussed in previous chapters can collect data in many formats. Managing infrastructure that collects data in every format is a cumbersome challenge and prone to failure and errors. The system architecture needs to detail which protocols are preferred and which can be accepted. For mature organizations, start with which protocols are currently in use and set end-of-life dates for any protocols the organization wishes to remove. It is strongly recommended to stick with default ports where they exist, and where applicable in the environment.

Another consideration is whether data will be stored locally, remotely, or both. Local storage adds management overhead and cost but may be a requirement in some environments. Having remote storage reduces management costs, but it can remove the option of using the metrics from an application to make environment choices. An example of this would be the Prometheus **HorizontalPodAutoscaler** (**HPA**). We'll discuss this in a little bit more detail in the *Management and automation* section. The authors have used short-lived, volatile local storage for such considerations in the past, while using a remote, third-party-provided infrastructure for long-lived storage, and such a setup works well.

OpenTelemetry offers several configurations. The following reference architectures are designed to give a starting point for readers who need to architect a system.

For instance, the simplest way to architect data collection is for each application to send data *directly* to the backends, like this:

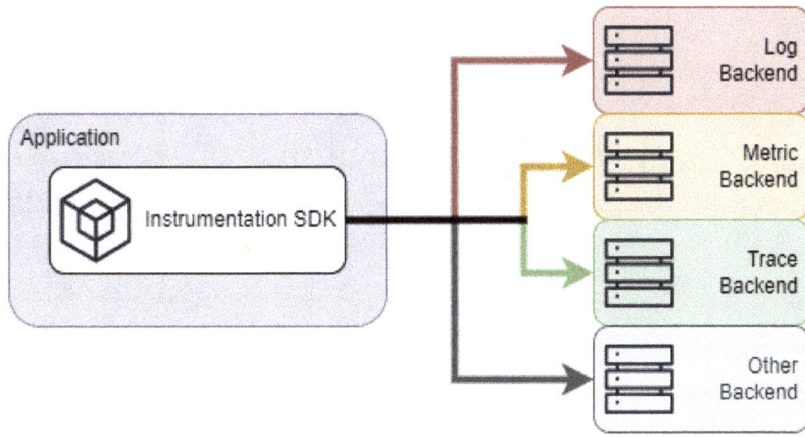

Figure 11.2 – Agentless configuration

For demonstrations or small installations, this architecture is perfectly fine. However, each application needs to be aware of each backend service, which means that this installation type does not scale very well.

Adding a **local agent** to the application adds a small amount of complexity but removes a lot of overhead for the team managing the application itself. Such an installation looks like this:

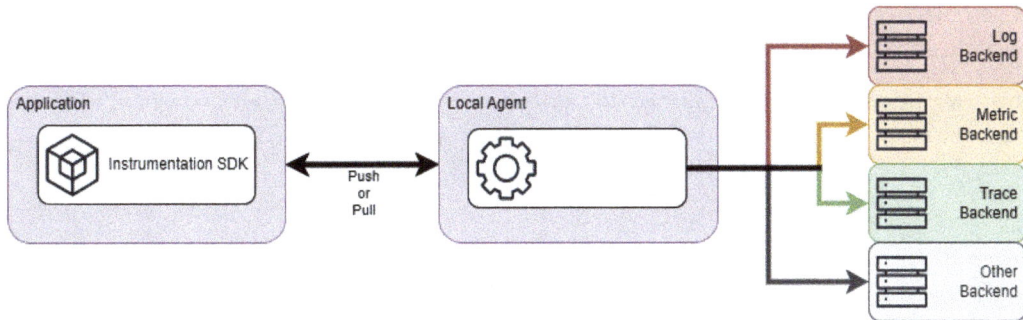

Figure 11.3 – Local agent only

Running a local agent is a very common pattern, and this pattern is great for many environments. As the number of instances of the agent grows, the agent configuration should be deployed using some form of configuration-as-code setup, such as Ansible, Salt, or Helm in a Kubernetes environment.

Adding a **gateway service** is another common architecture. This type of installation looks like this:

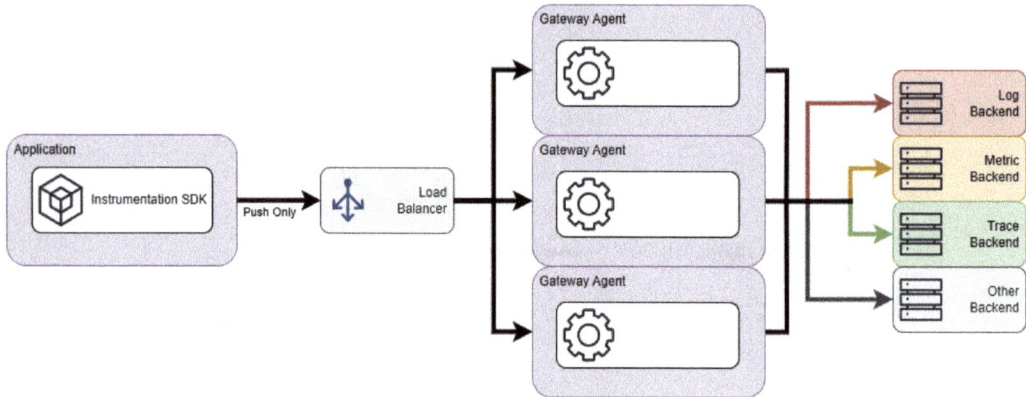

Figure 11.4 – Gateway agent only

Gateway architectures are perfect in a couple of situations:

- When the number of local instances is high, it can cause strain on the backend system by having a lot of open connections. Tthe gateway architecture resolves this by spreading this load over multiple instances of the agent.

- Gateway architectures are very good for installations where the collection of data from SNMP or similar systems is a goal.

It is best practice to put some form of load balancer in front of gateway architectures, and where possible to implement autoscaling.

Kubernetes introduces its own challenges to data collection architecture; this next diagram tries to capture the most common tools needed to collect data across the cluster and node:

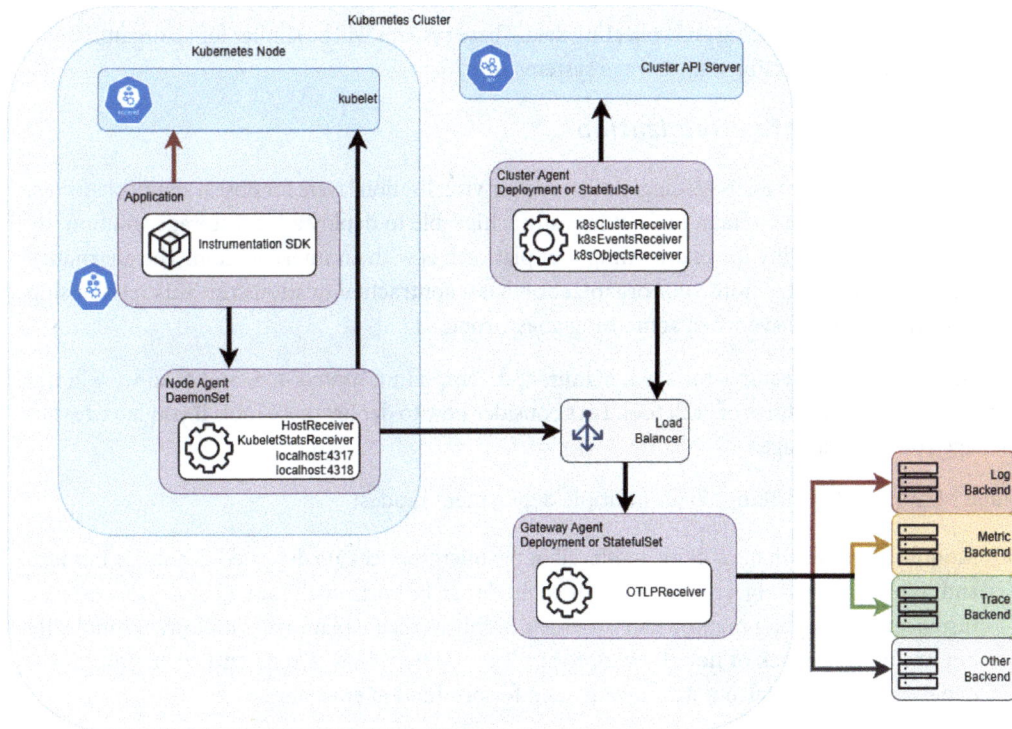

Figure 11.5 – A more complex Kubernetes architecture

A simple way to think of this configuration is to break it up into three parts:

- There is a **local agent** configuration on each node. This is configured to receive OTLP metrics on gRPC or HTTP. The local agent is also configured to query the kubelet for stats related to the Kubernetes node. It also has a host receiver configured to collect metrics; this would only be needed in a physical installation.

- A **gateway agent** is also configured. This collects data from each node and from the cluster agent. Using a gateway agent here also allows for processing to be done in the gateway.

- The final component is the **cluster agent**. This is a standalone instance of the agent configured to collect data from the Kubernetes API service. If this task were delegated to all node or gateway agents, the data would be collected by each instance, duplicating the data in the backend. By using a standalone instance, we get a single data stream, and this instance can leverage the gateway in the same way that the node agent can.

There are many more configurations that could be used, and we have not discussed the topic of using multiple different agents. However, this should give us a foundation to work from.

We have now looked at producing and collecting data. These systems will be similar for all organizations. Let's have a look at architecting data storage systems next.

Data storage and data visualization

There is one key question to ask regarding data storage or visualization layers for observability platforms. Who is responsible? With Grafana tools, it is easily achievable to deploy a local storage solution. By doing this, the responsibility for maintaining that platform is with an internal team. The alternative is to use a third party with whom your organization has a contractual relationship. This relationship is a very helpful thing to have when something goes wrong.

We considered the architectures for Loki, Mimir, and Tempo in *Chapters 4*, *5*, and *6*, so we will not show the specific architectures of each tool. Let's consider how to deploy these tools if you have reason to manage your own storage.

Grafana Mimir, Loki, and Tempo offer multiple deployment modes:

- **Monolithic mode**: In monolithic mode, all of the microservices are deployed as a single instance and connected to an object store. Monolithic mode can be horizontally scaled by deploying more instances. This scaling method can provide a highly available platform with lower complexity but has the drawback of not allowing for independent scaling of read and write paths. This deployment mode is also not recommended for production environments.

- **Microservices mode**: This mode deploys and scales each component of the system independently. This adds complexity but also allows the system to cater to the actual load that is placed on it. This mode is the recommended deployment mode for production use of Mimir and Tempo.

- **Simple scalable mode (only available in Loki)**: This mode strikes a balance between monolithic and microservices mode by allowing the independent deployment and scaling of write targets, read targets, and backend targets. These targets contain all the services needed for their role. This mode is the recommended deployment mode for the production use of Loki.

For all three of the storage platforms, deployment is carried out using a Helm chart for Kubernetes deployments. Packages are also supplied for deployments to Linux operating systems. These deployments can be automated using the provided Puppet or Tanka packages.

When you wish to manage your own data visualization layer, the Grafana application needs to be installed. This is available as a package for Linux, macOS, or Windows operating systems. Grafana also provides Docker images and detailed guidance on deploying to Kubernetes using Helm.

Handling system failure

A big consideration for a data collection architecture is how it handles failure. For agent failure, the only real option is to restart the agent. However, when the collection pipeline fails, this can be handled by buffering in memory or on disk. Each collector in the system is capable of buffering by configuring the batch processor for memory and the file storage extension for disk storage. The main thing to consider when designing a buffering solution is how long the system will need to tolerate failure. This, along with the throughput of data, dictates how much memory or disk space must be available to the instances. Reporting this calculation as a **service-level indicator** (**SLI**) for the data collection layer is a good practice, as it makes the resilience of the system publicly available.

We've now looked at how to architect the data that an observability system will collect, and we've looked at how to architect a system to collect that data. Let's now consider how to architect the system to account for management tools and automation tools.

Management and automation

We discussed using IaC in *Chapter 10*; when designing the system architecture, the use of IaC should manage the four systems (production, collection, storage, and visualization) as separate concerns:

- **Data production system**:
  - This should be managed by each application independently
  - Guidance should be provided
- **Data collection system**:
  - This is usually managed by an infrastructure, platform, observability, or similar team
  - This should have published SLIs and SLOs like any other component
- **Data storage system**:
  - This is usually managed by an infrastructure, platform, observability, or similar team
  - It is common to use a third-party tool (such as Grafana Cloud)
  - Grafana Cloud stacks are a great tool for separating storage where necessary, for example, for CI/CD platforms or performance testing
- **Data visualization system**:
  - The system itself would usually be managed by an infrastructure, platform, observability, or similar team
  - The dashboards and other artifacts related to an application should be managed by each application team independently
  - IaC can be provided to teams to manage deployment

Architecting for automation does not stop with the observability platform. Applications deployed to Kubernetes should be able to scale automatically as needed. When suggesting an ideal application pattern in the *Data production* section, keen-eyed readers may have seen that we recommended publishing metrics via a Prometheus endpoint as well as via OTLP export. This recommendation was made for autoscaling. While this book is concerned with observability in Grafana, a truly observable system can self-correct, such as the steam engine governor shown in *Chapter 1*. The Kubernetes HPA allows for the scaling of Pods based on CPU and memory usage. This is fine for some cases, but it is common for application teams to want to scale on metrics such as the rate of requests or number of sessions. The Prometheus community provides an adapter for Kubernetes Metrics APIs, which allows for querying a Prometheus endpoint to enable these types of scaling operations. An important question for an organization's architecture is whether this type of instrumentation would be managed by a central team or by each application team, perhaps with a default configuration offered for teams to consume.

We've looked at how to create an architectural design. There are a lot of tools available to test those designs in practice to prove they work. Let's have a look at proving the architecture.

Developing a proof of concept

The best place to prove a theoretical design is in an environment that has customers actually interacting with it, that is, a **production environment**. This is because any other environment is a mock environment and may miss some nuance of customer interactions. This is a recommendation to get the pathway to production created early and use it regularly. Having made that recommendation, it is still very important to have spaces for testing designs.

We will discuss compute containerization and virtualization tools, as well as simulated data production tools, which can be used to validate designs quickly.

Containerization and virtualization

Using containerization and virtualization locally and as part of a deployment pipeline can be a huge boost to provide quick feedback on whether a collection or storage architecture is achievable. Let's consider some of the tools that will help in this space:

- **Containerization**: The tools **k3d**, **KinD**, **MicroK8s**, and **minikube** can be used for containerization for the following reasons:

  - These four tools all offer the ability to run a Kubernetes cluster locally

  - KinD, k3d, and minikube can run using Docker or Podman drivers

  - minikube also offers a VM driver, which can be useful for certain local installations

  - For data collection architecture and pipelines, the authors have used KinD to deliver very good results

- **Virtualization**: **Vagrant** can be used with several virtualization tools, including **Hyper-V**, **VMware**, **VirtualBox**, **Xen**, **QEMU**, and **libvirt**. This is for the following reasons:

  - Vagrant offers the ability to define virtual machines and virtual networking and deploy these definitions on different virtualization tools using providers

  - This is a valuable feature for providing a reference virtual infrastructure for experimenting and use in a pipeline

These tools provide the capability to build reference infrastructure on which to deploy data collectors. They also provide the ability to document architectural requirements and diagrams using a real setup that is deployed locally.

Deploying infrastructure and data collectors is one part of the process of proving a design. Having tools to produce test data is also vital to check that the design is right. Let's have a look at these tools now.

Data production tools

There are a couple of ways of testing data production – using a sample application (such as the OpenTelemetry Demo application) or replaying pre-recorded datasets:

- **Demo applications**: These applications can be used to generate real observability data to test observability systems. Take the following examples:

  - **OpenTelemetry Demo application**: This is a full retail application that we have used to provide demo data throughout this book.

  - **One Observability Workshop applications**: These applications are provided by AWS and demo how to push data into AWS observability tools.

  - **mythical-creatures application**: This is an application written by Heds Simons for an interview with Grafana (he got the job). This application outputs metrics, logs, and traces. It's a simpler application than the OTEL demo, which can be an advantage.

- **Pre-recorded datasets**: These applications can be used to produce a predefined set of data to test observability systems. The process of replaying pre-recorded datasets crosses over very strongly with load-testing and packet capture tools. Tools such as **k6**, **Locust**, **Postman**, **Insomnia**, and **GHZ** can be used to send predefined data blobs to the data collection endpoints of your collection tools and validate the output. As observability tools use specific protocols, it's important to look for features that match the organization's production of data. Some examples are the following:

  - The ability to send gRPC data as this is a common format for OpenTelemetry

  - The ability to send other protocols such as SNMP if they are used

Tools such as **Fiddler** and **Wireshark**, as well as other network analyzers or HTTP(S) debuggers, can be used to record wire data to build up a library of reference data.

We will discuss in greater detail, in *Chapter 14*, how these tools can be integrated into CI/CD pipelines.

We've now seen how to architect the different components of an observability platform and how to validate those designs. Another important architectural consideration is getting the access levels correct. Let's look at that now.

Setting the right access levels

We have talked about the data in observability systems and how to architect the actual systems for producing, collecting, storing, and visualizing the data. A significant element of the architecture of the system that we have not discussed is RBAC.

There are two places where RBAC can be applied:

- **Grafana Cloud**: Administration of the deployed Grafana stacks and billing.
- **Grafana instances**: Access to data and visualizations. These instances can be deployed to Grafana Cloud or on-premises.

Let's start by looking at the permissions currently available in Grafana Cloud:

Permission/Role	Admin	Editor	Viewer
View API keys	✓	✗	✗
Manage API keys	✓	✗	✗
View organization billing information	✓	✓	✓
Manage organization billing information	✓	✗	✗
Manage Grafana Cloud subscription	✓	✗	✗
View Grafana instance plugins	✓	✓	✓
Manage Grafana instance plugins	✓	✗	✗
View stacks	✓	✓	✓
Manage stacks	✓	✓	✗
Manage organization members	✓	✗	✗
View invoices	✓	✓	✓
Pay invoices	✓	✗	✗
View Enterprise licenses	✓	✓	✓
View OAuth clients	✓	✓	✓

Manage OAuth clients	✓	✗	✗
View support tickets	✓	✓	✓
Open support tickets	✓	✓	✗

Table 11.1 – Grafana Cloud RBAC

These Grafana Cloud roles are focused on managing a Grafana Cloud instance. For most users, using and editing items in one or more Grafana instances is more applicable to their daily work. Grafana offers a rich permission set that breaks down into the following:

- **Basic roles**: The basic roles have very broad privileges. This is great for small organizations and having easy access to new installations. Assigning a basic role with least privilege to users is good practice. The basic roles are a default set of fixed role definitions, which we'll discuss in the next point. Basic roles include the following:

  - **Admin**: An admin for a Grafana organization.

  - **Editor**: A user who has access to edit objects in the organization.

  - **Viewer**: A user who has access to view objects.

  - **None**: A role that has minimal privileges for use with service accounts

  - **Grafana Admin**: A special admin account for all the Grafana organizations in an on-premises instance. As we have mainly discussed Grafana Cloud, let's clarify what a Grafana organization is:

    - Organizations are a method to separate Grafana resources in a single instance.

    - In Grafana Cloud, organizations are not available to use. Stacks are a better way to separate parts of the organization as a dedicated Grafana instance will be used in each stack.

- **Fixed role definitions**: Fixed roles can be used to expand the privileges assigned via basic roles. Fixed roles contain specific permission assignments that can be added to a subject.

- **Custom roles**: Custom roles allow for the creation of roles that have specific permissions, actions, and scopes assigned to them. Custom roles can only be created via the API, but Terraform can be used to manage these with IaC.

Permissions can also be assigned at the data source, team, dashboard, and folder levels. This can allow for structures such as giving management capabilities to all dashboards in a folder assigned to a specific team, but not granting management to other team folders. All the permission structures can also be managed using IaC, which we discussed in *Chapter 10*. Grafana provides a helpful guide on planning an RBAC rollout strategy here: `https://grafana.com/docs/grafana/latest/administration/roles-and-permissions/access-control/plan-rbac-rollout-strategy/`.

Let's consider how we might configure roles for some of the personas we have – *Diego Developer*, *Steven Service*, and *Pelé Product*:

- As a member of a team responsible for a service, *Diego* will need to be able to read dashboards to understand how other services may be behaving. He will also need to have write access for dashboards and alerts, but is limited to the folder that contains the application he is responsible for.

- *Steven* needs to be able to view dashboards but not edit them. However, he does need to be able to view and manage on-call schedules and silence alerts.

- *Pelé* has a couple of distinct needs. For most day-to-day processes, he needs to be able to view dashboards, incident history, and query data about the applications he is the product owner for. However, he also needs a service account to run specific queries for business metrics and load the data into the BI platform that is used with *Masha Manager* to analyze whether the teams need any help with delivering great products. He worked on setting up this service account with limited permission with *Ophelia*, the admin of the Grafana system.

For most users, once a role is created, it is simply a case of assigning the role to the individual user. Special consideration should be made for service accounts. Some service accounts, such as those used by the team managing the provisioning of Grafana tools, will need significant access and should be thoroughly audited. Other accounts, such as those used by an individual application team to manage dashboards, should have limited permissions. With this second type of account, it is a good idea to grant limited privileges for managing the service account to a senior member of the team as this enables the team to work independently.

Now that we understand RBAC in Grafana, let's have a look at how data collected for Grafana can be used in other systems.

Sending telemetry to other consumers

It is common for the data collected by observability systems to be of use in other systems. Logs are often used in SIEM systems and aggregate metrics are of interest in BI systems. There are two different strategies that can be used to share telemetry with other consumers:

- **Sharing data in the collection pipeline**: Sharing data in the collection pipeline is dependent on the data collection pipeline being used. We've talked a lot about the OpenTelemetry collector, which offers the ability to filter and send data to multiple backend systems. Similarly, AWS, GCP, and Azure offer options for writing telemetry to multiple backend systems. A consideration is that this type of solution will increase costs by storing multiple copies of the same data. Spending time with other consumers to understand their needs to minimize this cost is advised.

- **Querying data from Grafana directly**: Querying data from Grafana is done using a scheduled job that runs queries directly against Grafana. These are often custom connectors that will read data and write it into a BI platform. Grafana offers the recording rule functionality, which can assist in this data collection process. This functionality allows for the pre-computation of queries, which can be stored as a separate time series. For example, if the business were interested in the number of unique users who logged in daily, a recording rule could query this and store the data as a new metric. When the BI platform then collects this data, it would not need to wait for a potentially slow query to complete and instead would have the data easily available.

You should now be confident in architecting a comprehensive observability platform that meets the needs of the organization and can feed valuable information into other systems across the organization.

Summary

In this chapter, we have explored the process of architecting the data fields that will be collected. You will be able to use this knowledge to structure data in a Grafana platform so it is easy to use across your organization. We have discussed the process of architecting data production by applications and offering standard guidance on the best application structure to use. This will account for most needs of the developers in the organization. We shared several levels of complexity for the data collection architecture. You can use these as a starting point for architecting your own system. We discussed the various tools that are available to validate an architectural design: both tools for running local infrastructure and tools to simulate data that is being collected. This will help in producing a pipeline for delivering the infrastructure for an observability platform that you can rely on. Finally, we briefly discussed how to share data with other consumers, either in the data collection pipeline or by querying Grafana directly. You can use this knowledge to link observability data back to the rest of the organization.

In the next chapter, we will explore the use of **real user monitoring** (**RUM**) to collect data directly from the browser. This provides visibility of how your code runs when users are active in the system.

Part 4: Advanced Applications and Best Practices of Grafana

There are a number of topics related to observability, including frontend observability, application performance, load testing, DevOps pipelines, and monitoring security applications. This part will discuss these topics and additionally look at possible future trends. We will close out with some best practices and troubleshooting approaches.

This part has the following chapters:

- *Chapter 12, Real User Monitoring with Grafana*

- *Chapter 13, Application Performance with Grafana Pyroscope and k6*

- *Chapter 14, Supporting DevOps Processes with Observability*

- *Chapter 15, Troubleshooting, Implementing Best Practices, and More with Grafana*

12
Real User Monitoring with Grafana

In this chapter, we will investigate **real user monitoring** (**RUM**) with **Grafana Cloud Frontend Observability** and the **Faro Web SDK**. We'll explore what RUM is, how we use it to solve real user problems, and some important metrics to consider. We will then look at how to set up Frontend Observability in your Grafana Cloud instance and the prebuilt dashboards that are included. We will explore how Frontend Observability data can be correlated with backend telemetry for a more complete picture. To finish, we will look at best practices for collecting frontend data.

In this chapter, we are going to cover the following main topics:

- Introducing RUM
- Setting up Grafana Frontend Observability
- Exploring Web Vitals
- Pivoting from frontend to backend data
- Enhancements and custom configurations

Introducing RUM

RUM is the term used to describe the collection and processing of telemetry that describes the health of the frontend of your web applications. It gives us a bird's-eye view of user transactions as they happen, live from the user's browser all the way through to the backend system. The benefit of this telemetry is in the insight into the experience real users are having with the performance of your application.

Grafana implements RUM with a combination of the following:

- The Grafana Faro Web SDK, which, when embedded in your web application, collects the following telemetry by default:

  - Web Vitals performance metrics

  - Unhandled exceptions

  - Browser environment information

  - Page URL changes

  - Session identification (for data correlation)

  - Activity traces

  In addition to the defaults, the SDK can be configured to send custom metadata, measurements, and metrics into Grafana to enhance Frontend Observability. The Faro Web SDK integrates with opentelemetry-js to provide Open Telemetry-based tracing in your application. This SDK is open source, and the repository can be found at `https://github.com/grafana/faro-web-sdk`; it has comprehensive documentation along with some demonstration code.

- A cloud-hosted receiver of browser telemetry (you can alternatively deploy a Grafana Agent on your own infrastructure). The cloud-hosted receiver is created and configured when you activate application observability in Grafana Cloud (we will talk you through the setup steps in the next section).

- A dedicated Grafana app interface for each Frontend Observability app you create with included dashboards as tabs (we look at these dashboard tabs later in this chapter).

The following diagram shows the relationship between your *frontend application* in the left box, with the Faro Web SDK installed, and the *Grafana Cloud components* in the right box for ingesting, storing, and presenting your telemetry:

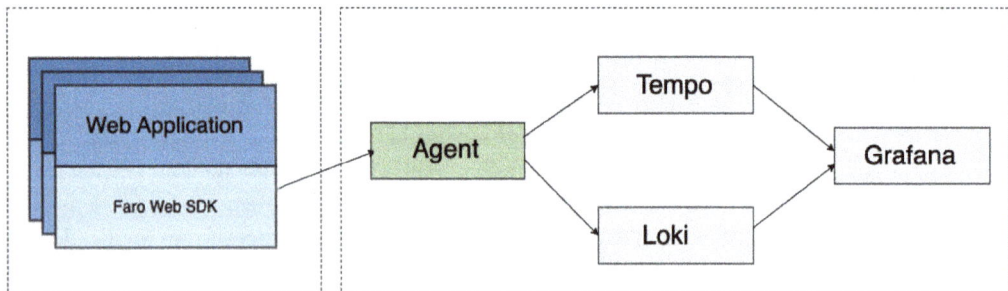

Figure 12.1 – Frontend relationship diagram

The Grafana Faro Web SDK collects telemetry and forwards it to the collector endpoint in Grafana Cloud where it is processed and sent to the appropriate backend – Loki for logs and Tempo for traces. Metrics are generated from logs using Loki's LogQL metric queries we discussed in *Chapter 4*. The generated metrics are visualized in the Grafana Cloud Frontend Observability application dashboards.

Let's now look at the setup for Grafana Frontend Observability.

Setting up Grafana Frontend Observability

To get started with monitoring your application frontend and configure Grafana Cloud Frontend Observability, follow these steps:

1. In your Grafana instance, select **Observability | Frontend** from the menu:

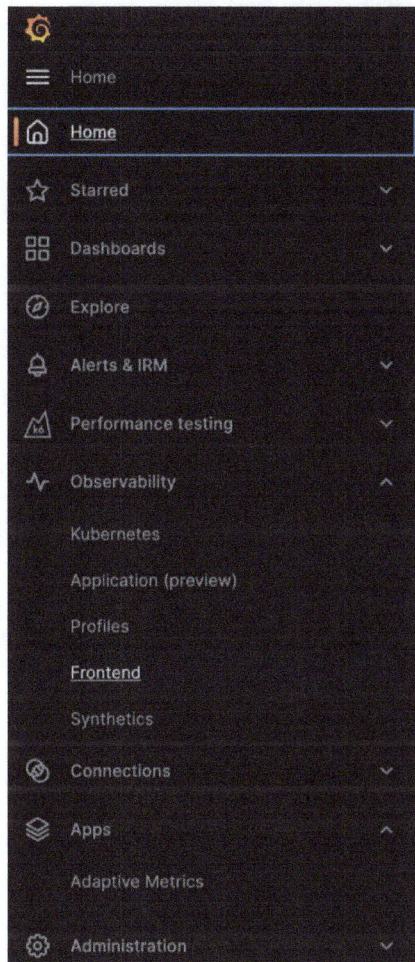

Figure 12.2 – Grafana Observability menu

2. The **Frontend Apps** landing page will be displayed. If this is your first time here, you will have a **Start observing** button – go ahead and click on it:

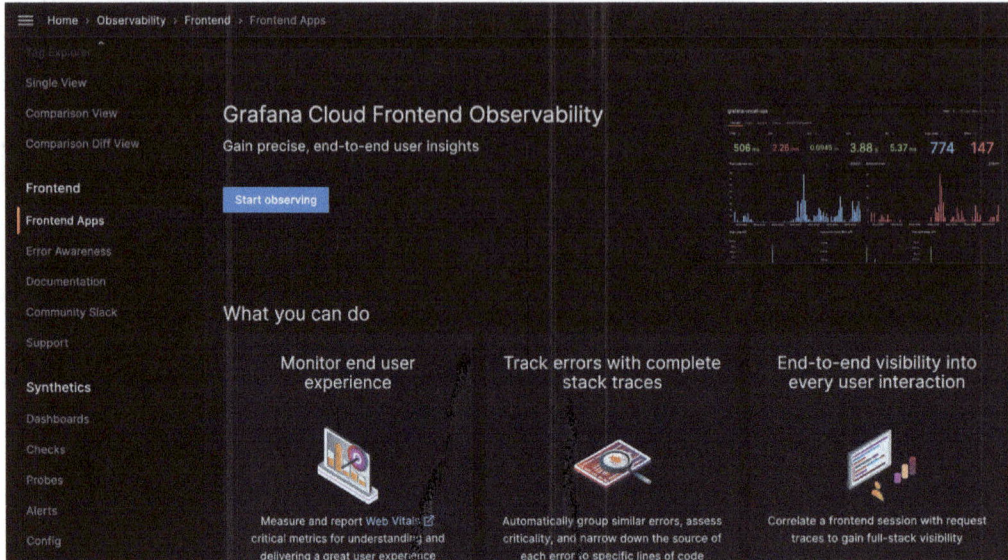

Figure 12.3 – Grafana Cloud Frontend Observability screen

If you have already set up frontend apps, they will be listed here and there will be a **Create New** button instead of the **Start observing** button. Clicking either of these buttons shows the **Create New App** modal window.

3. The **Create New App** modal window shown in *Figure 12.4* requires an app name, **Cross Origin Resource Sharing** (**CORS**) addresses, and additional label details (these are labels to add to Loki logs as they come into Grafana Cloud). Additionally, you will be asked to confirm the cloud costs associated with the additional telemetry from this feature:

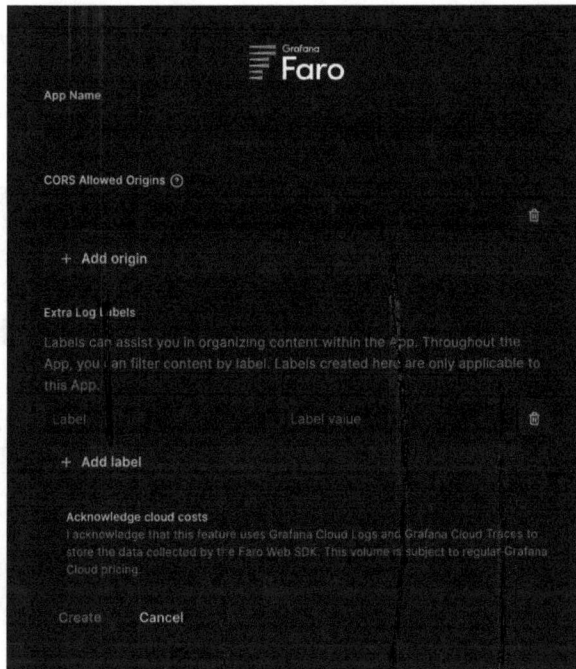

Figure 12.4 – Create New App screen

4.   Once you have completed the form, you are provided with a few options for integrating the Faro Web SDK into your frontend application using **NPM**, **CDN with Tracing**, or **CDN without Tracing**. You will have to decide which suits your application development requirements best:

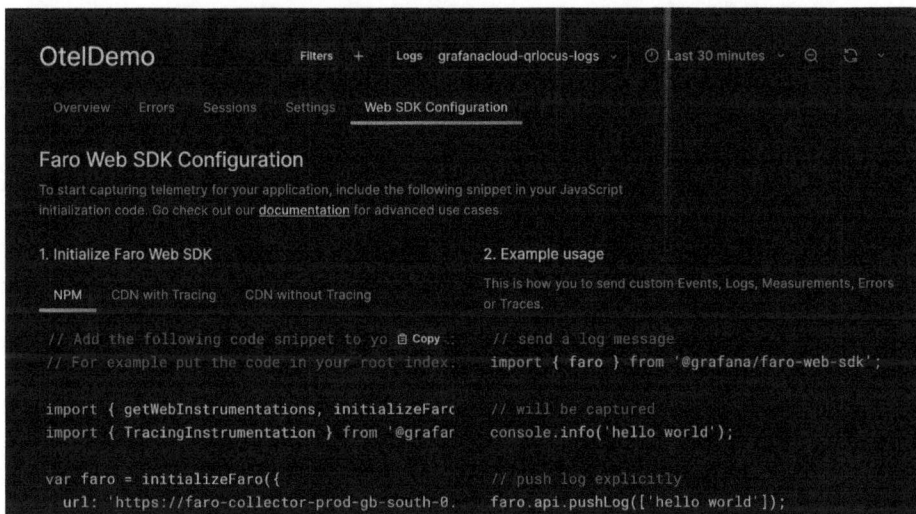

Figure 12. 5 – Web SDK configuration

When you have connected the Faro Web SDK in your application to Grafana, the default telemetry we identified at the beginning of this section will begin sending.

The **Overview** tab of your observability frontend app will look similar to the following screenshot, with the main dashboard included in Grafana Frontend Observability showing key metrics:

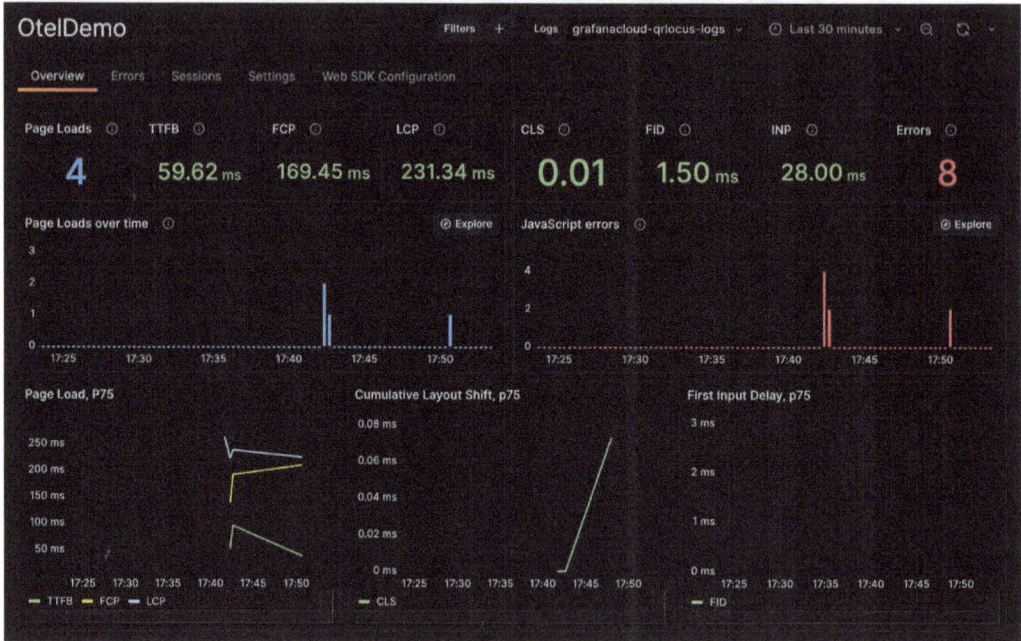

Figure 12.6 – Frontend Observability Overview tab

Next to the **Overview** tab are the **Errors** and **Sessions** tabs, which help you investigate your app. Let's look at these tabs in more detail:

- The top row on the dashboard in the **Overview** tab shows important **Web Vitals** metrics, including **Core Web Vitals**. Web Vitals is an initiative by Google that provides unified guidance for signals that are essential to reporting user experience on the web.

- The **Errors** tab details any frontend exceptions, where they happened, and the browsers affected:

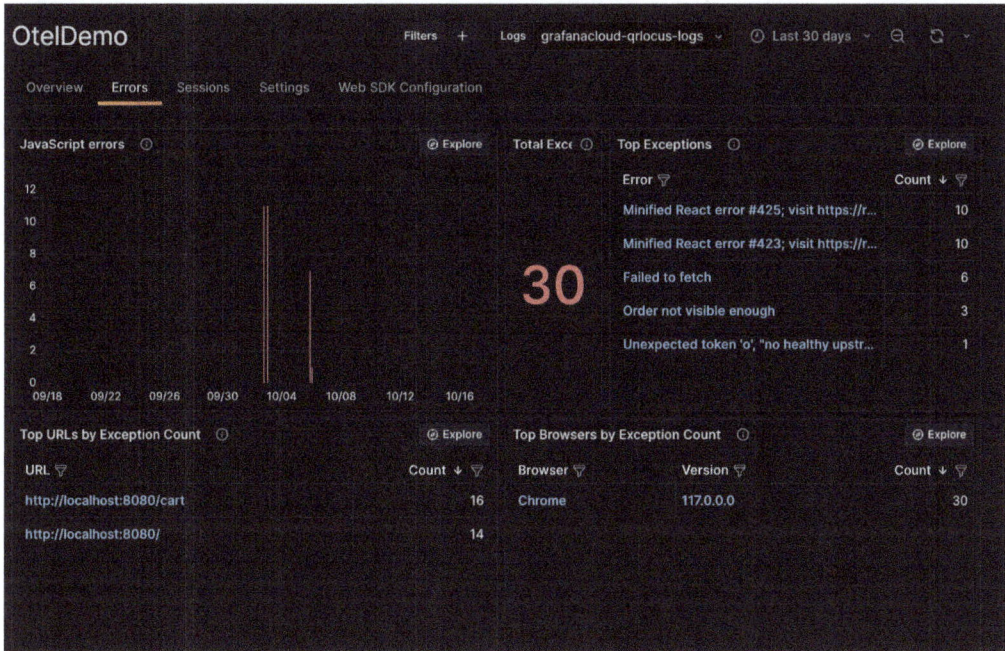

Figure 12.7 – Frontend Observability Errors tab

- The **Sessions** tab shows a list of the available user sessions to analyze. We will explore this in more detail in the *Pivoting from frontend to backend data* section.

The **Settings** and **Web SDK Configuration** tabs help you configure the connection (**Settings** allows you to modify the values entered in *step 3* earlier in this section, and **Web SDK Configuration** provides the config seen in *step 4*).

In the next section, let's look at Web Vitals metrics in more detail to understand the information we are capturing and reporting for our frontend applications.

Exploring Web Vitals

To explain the quality of experience delivered to users, the Web Vitals metrics report on several areas of a user's interaction. Web Vitals is a Google initiative that provides unified guidance for quality signals that are essential for delivering a great web user experience. You can read about the Web Vitals project in more detail at `https://web.dev/articles/vitals` and specifically Core Web Vitals at `https://web.dev/vitals/#core-web-vitals`.

The important Web Vitals metrics used in the **Overview** tab are as follows:

Metric	Description
Core Web Vitals	
Largest Contentful Paint (LCP)	The LCP metric measures the display time of the largest visible section in relation to the page starting to load. This could be text or an image that completely loads on the visitor's screen. A target LCP of <= 2.5 seconds is promoted. For more details, refer to `https://web.dev/articles/lcp`.
First Input Delay (FID)	The FID metric measures the time from when a visitor clicks a link to the time the browser starts processing the event. A target FID of <= 100 milliseconds is promoted. For more details, refer to `https://web.dev/articles/fid`. There are plans to replace FID with **Interaction To Next Paint (INP)** as a Core Web Vital in March 2024.
Cumulative Layout Shift (CLS)	The CLS metric measures any time a visible element changes its position. As the visitor experiences layout changes, the duration is captured and the cumulative score is reported. A target CLS of <= 0.1 is the aim. For more details, refer to `https://web.dev/articles/cls`.
Other Web Vitals	
Time To First Byte (TTFB)	The TTFB metric measures the time between the request for a web resource and when the very first byte of a response starts to arrive. This will not be obvious to the visitor, but it is a good indicator of responsiveness of getting content to the visitor. A target TTFB of <= 0.8 seconds is the objective. For more details, refer to `https://web.dev/articles/ttfb`.

First Contentful Paint (FCP)	The FCP metric measures the time from when the page starts loading to when any part of the page is displayed on the screen – essentially, when your visitor actually sees an interaction with the website requested.
	A target FCP of <= 1.8 seconds is the aim.
	For more details, refer to `https://web.dev/articles/fcp`.
Interaction to Next Paint (INP)	The INP metric measures the responsiveness of a page throughout a visitor's session. It does this by observing the latency of every click, tap, and keyboard interaction by a visitor and takes the longest (ignoring outliers).
	The target value for the INP is <= 200 milliseconds.
	For more details, refer to `https://web.dev/articles/inp`.

Table 12.1 – Important Web Vitals metrics

Now that we have looked at some of the important frontend metrics to consider, let's look at how we can pivot to backend telemetry when investigating issues.

Pivoting from frontend to backend data

Once you have started collecting Frontend Observability data, you will be able to correlate it with backend and infrastructure telemetry. Grafana provides simple interfaces for this when you are using Loki for logs, Tempo for traces, and Mimir for metrics.

Within the Grafana Cloud Frontend Observability app, there are readymade dashboards that make navigation and investigation simple. As discussed in the *Setting up GrafanaFrontend Observability* section, the app menu has three main sections, namely, **Overview**, **Errors**, and **Sessions**. The **Sessions** tab allows us to jump into other telemetry that our system is producing and sending to Grafana. You can see in the following screenshot multiple requests to different page URLs alongside the associated session IDs:

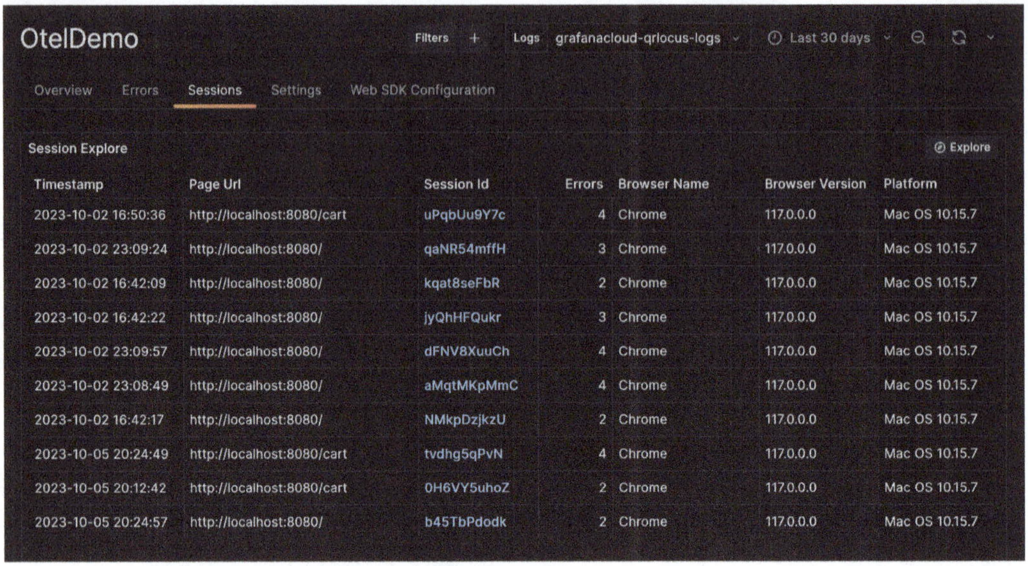

Figure 12.8 – The Sessions tab

Clicking a **Session Id** entry will take you to a detailed view, where you can see the Web Vitals for that specific visitor session, associated exceptions, and, if you have instrumented the rest of your system, links to **Traces**:

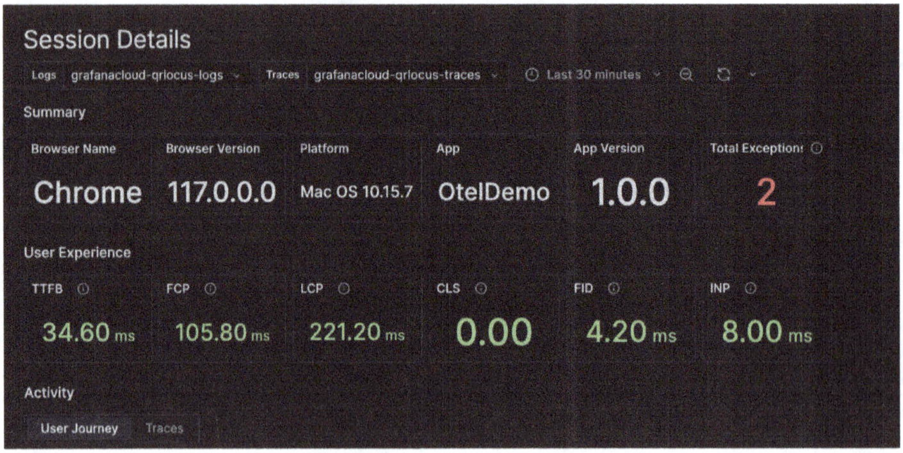

Figure 12.9 – The Session Details screen

Selecting **Traces** takes you into an **Explore** view of full system traces, where you can easily navigate to associated Loki logs from the backend, providing you with the ability to traverse your system. This demonstrates the additional value that can be gained with full end-to-end observability instrumentation:

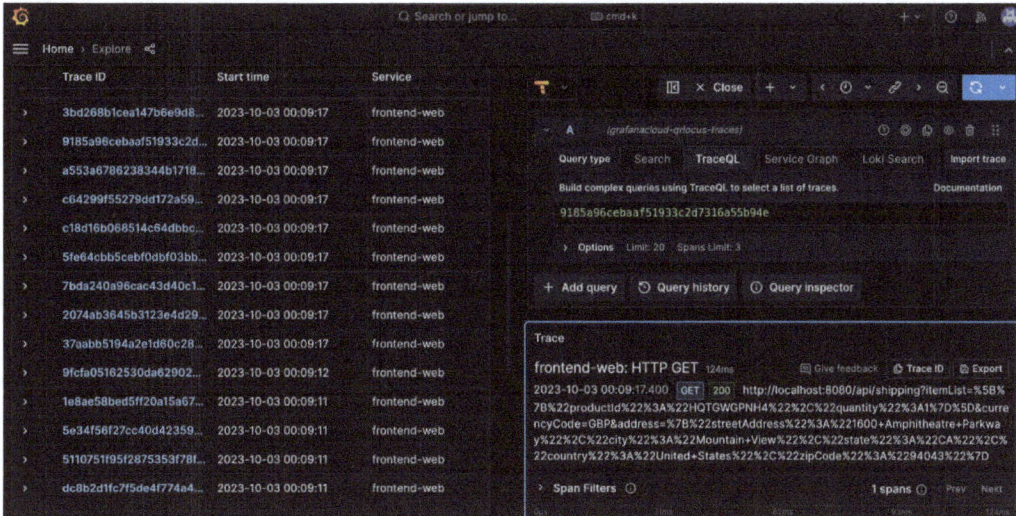

Figure 12.10 – The Explore view for system traces

A default session is defined by the following:

- **Session start**: The visitor navigates to a web page, Faro is initialized, and a new session is started
- **Session end**: The visitor navigates away and the session is destroyed (by default, a session end event is not sent)

However, you can define session logic as you prefer, to fit your use case. The Faro docs for session instrumentation at `https://grafana.com/docs/grafana-cloud/monitor-applications/frontend-observability/faro-web-sdk/components/provided-instrumentations/#session-tracking` show available configurations.

Let's now look at some best practices to consider with Frontend Observability.

Enhancements and custom configurations

With all observability, you need to consider your use case. This is especially important when considering Frontend Observability as you will be operating in your visitor's browser. There are several enhancements you can make to Frontend Observability over the default implementation. However, these enhancements come with the overhead of additional configuration, additional developer effort, and a potentially greater impact on your visitor's browser. However, they can dramatically increase the value provided by Frontend Observability instrumentation.

Let's explore the enhancements for Frontend Observability:

- **Frontend Tracing**: This provides improved correlation between real user interactions and backend events. It requires the additional OpenTelemetry SDK configurations and adds some overhead to your visitor's browser, so consider the implications carefully and test for any impact.

- **Custom Errors**: This provides improved observability for systems that have error handling. Additional configuration is required to manually add the Faro Web SDK to the error handling pipeline to push errors to Grafana.

- **Custom Measurements**: This provides enhanced telemetry with application-specific data. Additional configuration is required to manually add the Faro Web SDK to push additional measurements to Grafana.

- **Custom Logs**: This provides the ability to send supporting metadata along with telemetry to help you understand your visitors' experiences. It requires additional configuration and development effort to instrument.

- **Custom Events**: This provides additional correlation to help with issue investigation and data presentation. Events are ingested by Loki as logs with a specific label, `kind=event`. We learned all about Loki labels in *Chapter 4*.

If you do not have a frontend app ready to instrument, there is a demo project as part of the Faro Web SDK that you can experiment with. It will help you understand in more detail what is happening and how it works: `https://github.com/grafana/faro-web-sdk/blob/main/demo/README.md`.

Let's now wrap up this chapter with a summary and look toward the following chapter.

Summary

In this chapter, we have explored user monitoring and the additional value gained with full end-to-end observability instrumentation. We looked at how Grafana provides this capability with Grafana Frontend Observability and the Faro Web SDK. We then looked at the Web Vitals metrics, which are important for interpreting visitor experience. We also looked at the built-in dashboards that help you navigate to your backend telemetry in Grafana, giving you the ability to fully diagnose problems. Finally, we looked at some best practices and custom configurations for Frontend Observability.

In the next chapter, we will learn about different aspects of application performance with **Pyroscope** and **k6** from Grafana.

13

Application Performance with Grafana Pyroscope and k6

This chapter will explore two tools, **Pyroscope** and **k6**. Pyroscope is a **continuous profiling** tool that allows users to collect very detailed information about the usage of system resources such as CPU and memory. k6 is a **load testing** tool that can be used to interact with an application via endpoints, or via a browser session in a scripted way.

With Pyroscope, we will see how to search data, which will give you a good understanding of how to make use of the data available. We will then show how to add instrumentation to collect this data using both an installed client and by adding a native language SDK to the application code. Finally, we will see how the new version of the Pyroscope architecture leverages Grafana's knowledge of highly scalable storage platforms, using inexpensive block storage to set Pyroscope on a path toward offering truly continuous profiling for developers. This functionality will allow those of you who need visibility of code execution to improve operational cost or end user performance.

k6 will move a little away from observability into the very closely related field of load or performance testing. We will discuss the general principles of load testing and look at the different categories of load tests that you may need. Then, you will be introduced to the scripting language used by k6 to easily write tests that validate the application is performing as expected. We will see how k6 uses **virtual users (VUs)** to scale tests and create a significant load on an application, so you can use it to prove your applications are running as expected. Finally, we'll see how k6 can be installed, and how it is versatile enough to even run as part of a CI pipeline, ensuring your applications are continuously load tested.

In this chapter, we're going to cover the following main topics:

- Using Pyroscope for continuous profiling
- Using k6 for load testing

Using Pyroscope for continuous profiling

First, let's address the question of what **continuous profiling** is. As we outlined at the start of this book, a system is observable when the internal state of the system can be inferred from its external outputs. We have seen three types of output telemetry: logs, metrics, and traces. Profiling data is another form of telemetry. Profiling data is very low-level data that relates to a workload's use of resources, such as the use of CPU or memory. As profiling tools analyze very low-level system data, they capture information such as the running time or the number of objects in memory of a specific application function. This is very powerful for domain experts to inspect how an application behaves, and this power can lead to significant performance and cost improvements. Profiling has been around for a long time, as anyone who has produced a stack trace will know. Pyroscope offers the ability to capture this profiling data continuously, with a default interval of 15 seconds. The ability to collect this telemetry continuously over the lifetime of an application can give insight into how an application runs over time, which can link the inner workings of the code base to specific user actions seen in logs, metrics, and traces.

In this section, we will briefly introduce Pyroscope. You will be shown how to search the data collected by Pyroscope. We will talk about configuring the client to collect profiles, and we will look at the architecture of the Pyroscope server.

A brief overview of Pyroscope

Pyroscope, also known as Grafana Cloud Profiles, was founded in 2020 and acquired by Grafana Labs in 2023. The Pyroscope team joined the team from a Grafana Labs experimental product called Phlare, and the product is now a standard offering from Grafana Cloud. Some of the key features of Pyroscope are as follows:

- Great horizontal scalability using the same architecture as Loki, Mimir, and Tempo
- Cheap storage for profile data
- Can store data locally or using Grafana Cloud
- High frequency of sampling, which produces very granular data

Now, let's explore how we can examine the data collected by Pyroscope.

Searching Pyroscope data

Profile telemetry can be viewed using the **Explore** view in the Grafana UI by selecting a Pyroscope source. While the view is similar to Loki, Mimir, and Tempo, the query language is limited by the nature of the telemetry type; effectively, only selection functionality is available to select a signal from an application or group of applications by tag. This is the view you will see to select data from a Pyroscope source:

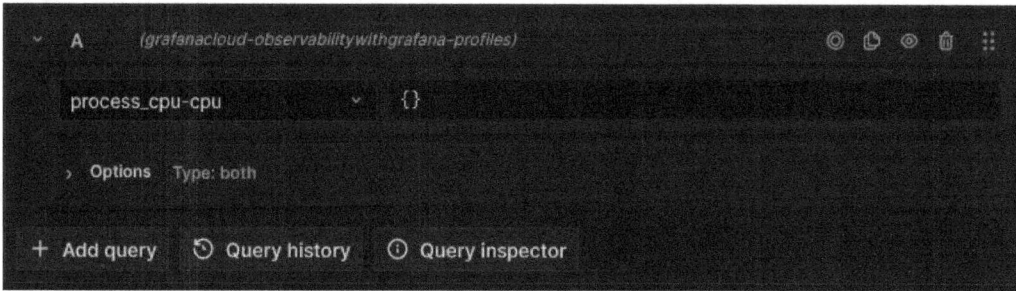

Figure 13.1 – Pyroscope query pane

The first view to look at is the **Top Table** view, as shown in *Figure 13.2*. For those of you who know the Linux `top` command, this view will be familiar. The view lists every function and the amount of time that has been spent on the function. The **Self** column shows the time spent on that function. The **Total** column shows the total time each function takes to run. This allows users to see functions that have a long running time. Long runtimes could indicate an inefficient function, but it could also indicate a function that is central to the application. Domain expertise is needed to understand where improvements could be made. This screenshot shows the **Top Table** view:

Symbol	Self ↓	Total
[unknown]	320 ms	7.52 s
do_epoll_ctl	113 ms	124 ms
try_to_wake_up	51.5 ms	51.5 ms
__wake_up_common_lock	41.2 ms	41.2 ms
memcg_stat_show	30.9 ms	30.9 ms
finish_task_switch	30.9 ms	30.9 ms
hrtimer_start_range_ns	30.9 ms	30.9 ms
page_reporting_process	20.6 ms	30.9 ms
kmem_cache_alloc	20.6 ms	20.6 ms
_raw_spin_lock	20.6 ms	20.6 ms
futex_wake	20.6 ms	72.2 ms
hv_free_page_report	10.3 ms	10.3 ms
__siphash_unaligned	10.3 ms	10.3 ms
__cond_resched	10.3 ms	10.3 ms
gc_worker	10.3 ms	41.2 ms
si_meminfo	10.3 ms	10.3 ms

Figure 13.2 – Pyroscope Top Table view

The second view is the **Flame Graph** view. This chart is specifically designed to visualize profile data. **Flame graphs** were invented to be able to visualize stack trace output from applications to make debugging easier. Before we look at the view in Pyroscope, let's take a look at a sample application stack trace:

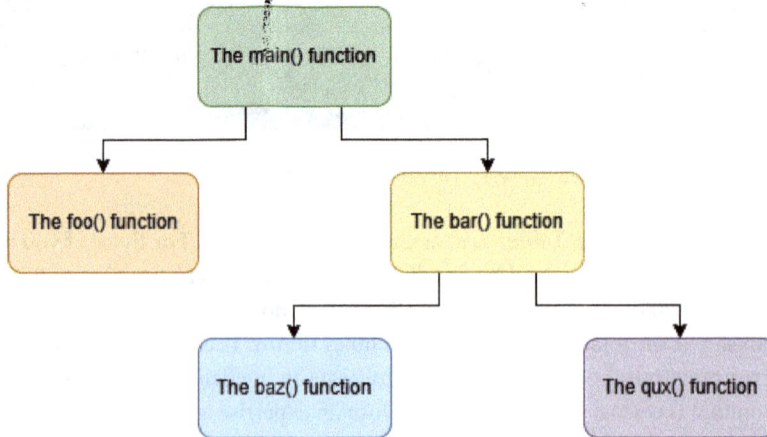

Figure 13.3 – Example application stack trace

We can see there is a `main()` function, which is started when the application is run. This function calls the child functions, `foo()` and `bar()`, in order, and `bar()` also calls the `baz()` and `qux()` functions. A flame graph captures the hierarchical nature of these stack calls by grouping child functions under their parent. This allows us to see how deep the call stack is by looking at the y axis. The total population of functions is shown in the x axis; importantly, this does not represent the time but rather each function that was seen on the call stack during the sampling period. The visualization of duration is shown in a flame graph by the width of the box for each function, which shows the total time spent on a function during the sampling period. Let's have a look at how this looks in practice:

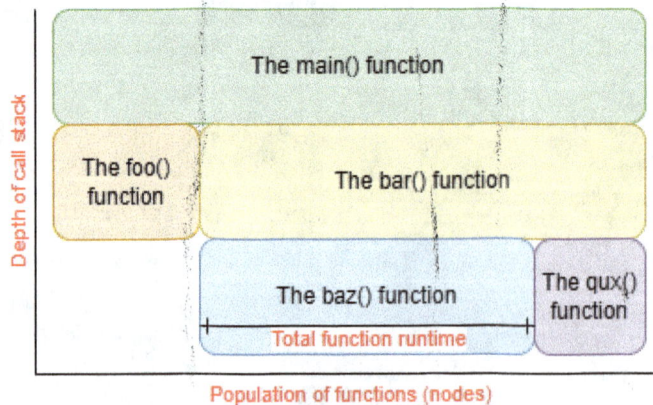

Figure 13.4 – Example flame graph from the stack trace

In this example flame graph, we can see that the baz() function takes up a significant portion of the operating time. In some applications, this may be completely expected behavior; in other applications, this may indicate a function that needs to be optimized.

Very few applications are as simple as this example. Let's look at a real flame graph from the OpenTelemetry Demo application:

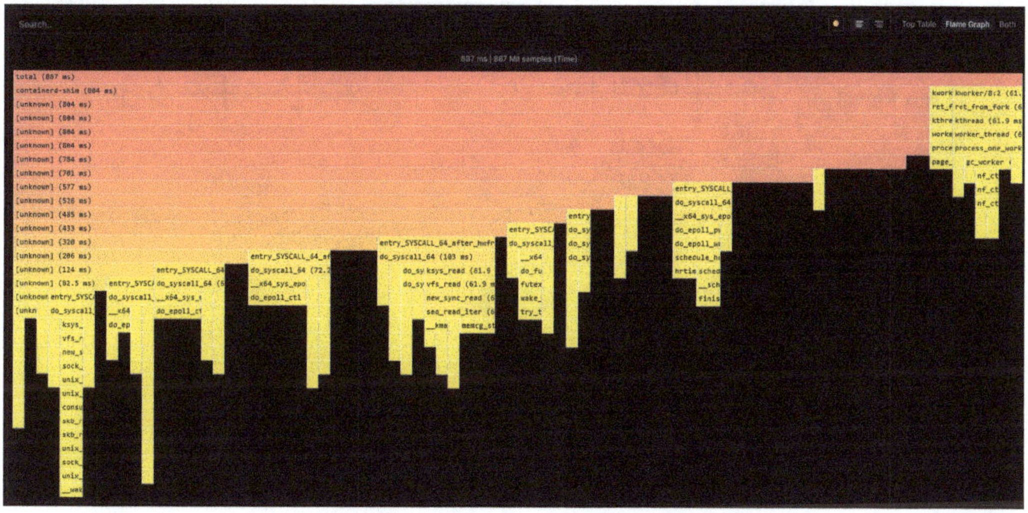

Figure 13.5 – A real flame graph

We've seen how continuous profiling tools such as Pyroscope can be valuable in creating efficient code and debugging issues. Let's now look at how to collect profile data.

Continuous profiling client configuration

There are currently three separate ways to collect data for Pyroscope, although we expect this to evolve as Pyroscope is quite a new piece of technology. We would recommend the Grafana Labs blog for those of you who want to keep up to date with the latest developments from this exciting technology (https://grafana.com/blog/). Let's explore how to set up each one:

- **Extended Berkeley Packet Filter (eBPF) client**: The first way to collect profile data for Pyroscope is to make use of a Linux kernel-level tool called eBPF. This tool allows the profiling client to view the trace information for all applications running on the server or node. The eBPF client combines this data with metadata on the data source (for example, a Kubernetes Pod or namespace) and then sends this profile information to a Pyroscope backend. The following diagram shows a simplified view of how eBPF stores data for the Pyroscope client to collect:

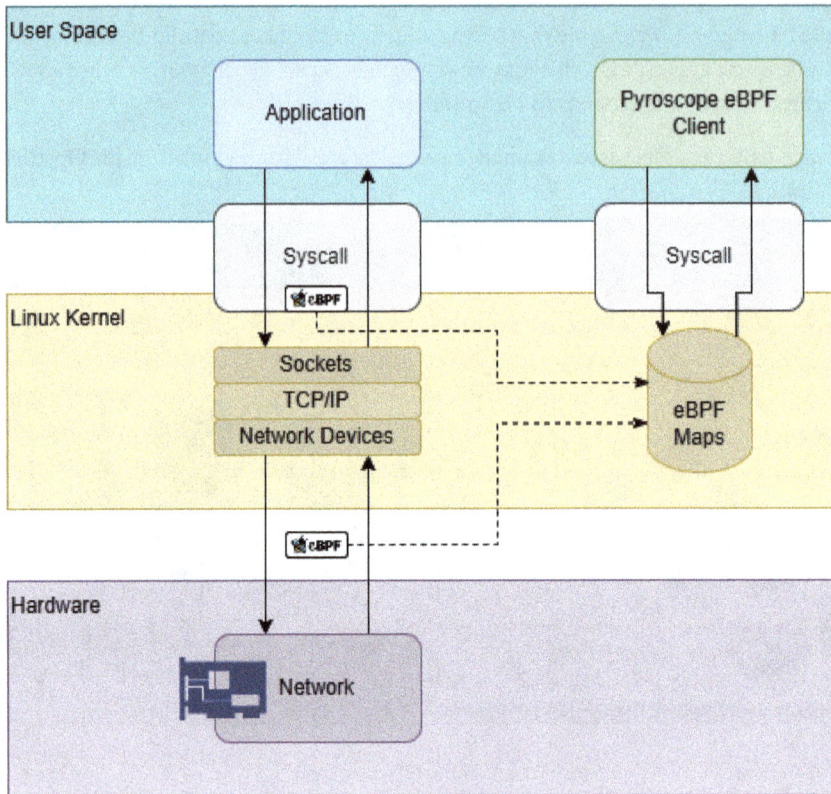

Figure 13.6 – eBPF client process

With eBPF, the kernel collects profile data, as well as several other types of data, and stores it in eBPF maps. Pyroscope links into the eBPF maps, packages the data, and then sends it for storage in the configured backend.

- **Native language instrumentation**: The second way to collect profile data is to use a language-specific Pyroscope SDK to add instrumentation to your application. SDKs are currently provided for Go, Java, .NET, Python, Ruby, Rust, and Node.js. Apart from the Go SDK, all these libraries only support a *push* mode of operation. Go supports both a *push* and a *pull* mode of operation; the pull mode allows the Grafana agent to collect profile data from a scraping endpoint published by the application. In push mode, it is currently necessary to add the Pyroscope server address, basic authenticated username, and password at the application level, although as this tool matures, we're sure this will become easier to manage in an operational environment.

- **Instrumenting Lambda functions**: Pyroscope also provides tooling for AWS Lambda functions. This consists of a Lambda extension that is loaded as a layer when the function is triggered. This allows the profiling tooling to collect the required profile telemetry asynchronously without impacting the operation of your Lambda function. Like the native language instrumentation, environment variables must be provided with the remote address for the Pyroscope backend and the relevant authentication tokens.

For teams running serverless functions, this adds the capability to look inside the Lambda function black box and allows teams to answer questions such as, *Why is my Lambda costing so much?*, *Why do I have such high latency?*, and *Why is my function failing so often?*

There are benefits and drawbacks to the eBPF client, the SDK, and the Lambda approaches: their usage is dependent on the use case. Here are some benefits and drawbacks for each method:

Instrumentation method	Benefits	Drawbacks
eBPF	System-wide whole system profiles are easy to collect. Infrastructure metadata is easy to add (for example, Kubernetes Pod or namespace). Easy to manage a multi-language or large system. Can combine with native language instrumentation.	Linux kernel constraints. Limited ability to tag user-level code. Some profile types are not performant to collect (for example, memory use). More complex for local development environments.
Native language	Flexible tagging of code. Detailed profiling of specific parts of code. Ability to profile other types of data (for example, memory use). Simple to use in local development environments.	Managing a multi-language or large system is difficult. Difficult to auto-tag infrastructure metadata (for example, Kubernetes Pod or namespace).
Lambda functions	Allows for collection of trace data from serverless functions. Links with the native language support to instrument the function.	Currently only available for AWS Lambda.

Table 13.1 – Advantages and drawbacks of Pyroscope instrumentation methods

We've looked at the different ways to set up applications and clients to collect profile data. Now, let's consider the storage and search architecture of Pyroscope.

Understanding the Pyroscope architecture

Pyroscope 1.0 has introduced a major change to the architecture of Pyroscope. This leverages the Grafana knowledge of Cortex architectures to make the architecture horizontally scalable. This is a breaking change from previous versions so we will only be considering the architecture from this change onward.

Similar to Loki, Mimir, and Tempo, Pyroscope uses low-cost, highly available block storage such as Amazon S3, Google Cloud Storage, or Microsoft Azure Storage to provide massive scalability. Here's a diagram of the Pyroscope architecture:

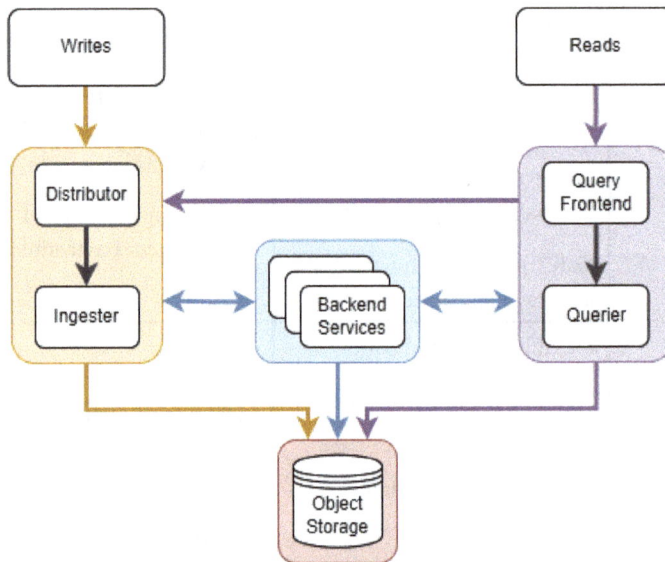

Figure 13.7 – Pyroscope architecture

When data is written, it is sent to the **Ingester**, which persists the data to **Object Storage**. On the **Reads** side, queries are split and sharded to instances of the **Querier**, which grabs the necessary data from the **Ingester** and/or the long-term storage.

There are several alternatives to Pyroscope on the market that may be of interest to you. The open source tools include OpenTelemetry eBPF, Parca, and profefe, and several observability vendors include similar profiling tools. These tools can be found at https://github.com/open-telemetry/opentelemetry-ebpf, https://www.parca.dev/, and https://github.com/profefe/profefe. We've now seen how Pyroscope functions. Another tool that is helpful for developers and testers is k6 load testing. Let's take a look at this next.

Using k6 for load testing

Load testing is the practice of applying a known, artificial load to an application to see how it behaves. The term is often used interchangeably with performance testing, and we will follow the k6 documentation in using *average load* to differentiate a specific type of test.

Several different types of load tests can be applied; they differ on two axes – the load throughput and the duration. They may also differ in the content of the tests that are performed. Some common types of tests are shown in the following table:

Test	Description	Purpose	Runtime and volume
Smoke tests	These are designed to validate that the system works. They can also be known as sanity or confidence tests. They are called smoke tests after testing a device by powering it on and checking for smoke.	These are designed to quickly say that things look as expected or that something is wrong	These should run quickly, in minutes not hours. They should be low volume.
Average load tests	These tests show how the system is used in most conditions.	These are designed to simulate the most frequent level of load on the system.	These should run relatively quickly, but slower than smoke tests. They should simulate average volumes of traffic.
Stress tests	These tests stress the system with higher-than-average peak traffic.	These are designed to simulate what would happen if peak traffic were experienced for an extended duration.	These should run in less than a day. They should simulate high volumes of traffic.
Spike tests	These tests should show how the system behaves with a sudden, short, massive increase in traffic, as might be seen during a denial of service (DoS) attack.	These are designed to test how the system would handle a sudden overwhelming spike in traffic, such as a DoS attack.	These should run quickly. They should simulate unrealistic amounts of traffic.

Breakpoint tests	These tests gradually increase traffic until the system breaks down.	These are designed to understand when the system will fail with added load.	These can run for extended periods. They should simulate steadily increasing rates of traffic.
Soak tests	These tests assess the performance of the system over extended periods. They are like an average load test over a significantly longer period.	These are designed to demonstrate how the system will function during real operations for extended periods. They are good for identifying issues such as memory leaks.	These will run over extended periods such as 48 hours. They should simulate average volumes of traffic.

Table 13.2 – Types of load tests

The following graph shows the different tests for reference:

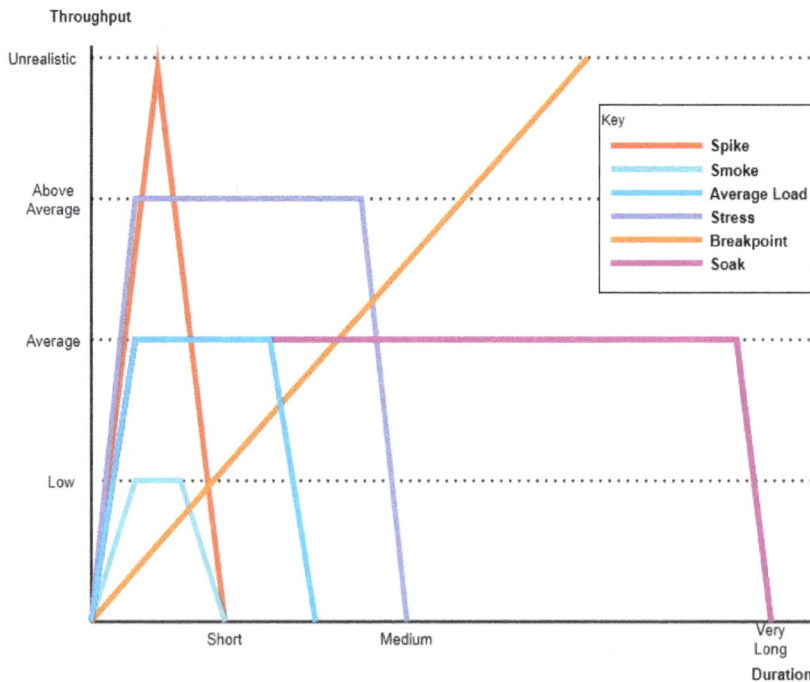

Figure 13.8 – Visual representation of the different load test types

In the preceding figure, we can see the different types of tests graphed by the test throughput and the test duration. Try correlating what you can see in the graph with what you've just learned about these tests in *Table 13.2*.

You can see that load testing and observability are very closely linked. The data collected from a live system will show what average and unrealistic loads look like. The data injected by a smoke test can show a system is working as expected, for example, after a new version is deployed. The data collected from the load testing environment can give critical insights into the operation of the system under load.

It is good practice to separate the observability data collected from load testing from other data. Due to the nature of the tests that are being tried, very large volumes of data can be generated, which can be a very costly thing to collect. One huge advantage of open source systems such as Grafana is the ability to run the data storage system as part of the load testing environment while using the same visualization as in production.

There are several load testing tools on the market, both open source and commercial. The open source offerings include JMeter, k6, Gatling, Locust, Artillery, Tsung, Vegeta, Hey, and Siege. As this book focuses on Grafana tools, we will only discuss k6 here. Let's have a look at some of the features of k6.

A brief overview of k6

k6 is the load testing tool developed by Grafana Labs after they acquired LoadImpact. k6 offers several key features:

- A **command-line interface (CLI)** that allows tests to be run, paused, resumed, or scaled.
- The ability to start tests locally, from a Kubernetes cluster, or in the cloud with the CLI. k6 supports distributed running via a Kubernetes operator.
- Scripting support using JavaScript.
- The ability to load additional modules into scripts, although this does not include support for Node.js modules.
- A browser module that adds browser-level APIs for full frontend testing.
- Support for goal-oriented load testing using checks and thresholds.
- Great supporting tools, such as the following:
  - Reference projects
  - Tools to convert scripts from other tools to k6
  - Tools to convert k6 output to other common formats
  - A GUI for test building

Let's now look at the process of writing a simple test.

> **Important note**
>
> As k6 requires a test file to run, we have included the installation and usage instructions after these instructions on writing a test.

Writing a test using checks

Tests are written in k6 using JavaScript. A very simple test to submit a GET request to the acme website would look like this:

```
import http from 'k6/http';
export default function () {
  http.get('http://www.acme.com');
}
```

This script would just submit a request to the web page, but it would not validate that the request was successful. The check functionality would be used to confirm that this is the case, like this:

```
import { check } from 'k6';
import http from 'k6/http';

export default function () {
  const res = http.get('http://www.acme.com');
  check(res, {
    'is status 200': (r) => r.status === 200,
  },
  { company: 'Acme' }
  );
}
```

The check() function takes a value, an object containing the checks that will be run against the value, and an object containing any tags. If all the checks pass, then the function returns true; otherwise, it will return false. The check functionality makes it very simple to check for simple conditions in a script. It is common to want to check that an endpoint is meeting specific expectations, and k6 offers thresholds for this goal.

Writing a test using thresholds

Thresholds are checked against all requests made in the script, and it is good practice to use the **service-level objectives** (**SLOs**) set by the team as a starting point for testing. Here is an example of a threshold test:

```
import http from 'k6/http';
export const options = {
```

```
  thresholds: {
    http_req_failed: ['rate<0.01'], // http errors should be less than
1%
    http_req_duration: ['p(95)<200'], // 95% of requests should be
below 200ms
  },
};

export default function () {
  http.get('http://www.acme.com');
}
```

This test would make a call to the acme website and check that the built-in http_req_failed and http_req_duration HTTP metrics meet the threshold expression specified. These metrics are collected from all the requests made in the script; in this case, there is only a single request made. If needed, it is possible to use **groups** and **tags** to evaluate HTTP requests independently.

Now that we know how to write basic scripted tests, let's look at how we can use options to scale.

Adding scenarios to a test to run at scale

In the previous section, we mentioned that the test would only make a single HTTP request. By using options, it is easy to manage the behavior of the default function in complex ways. Let's consider a simple example in which we create 100 VUs, and each VU will execute the default function repeatedly for 30 minutes:

```
import http from 'k6/http';

export const options = {
  vus: 100,
  duration: '30m'
};
export default function () {
  http.get('http://www.acme.com');
}
```

You might notice that we are using the same options constant as we used when we created the test thresholds in the previous section. The options configuration option offers a lot of flexibility for defining the behavior of a test. It is a common requirement to share data with each of the VUs that will run the tests. Let's have a look at how the test life cycle can manage these requirements.

Test life cycle

There are four stages to the k6 test life cycle. These stages are explicitly set in the ordering of a test file:

1. **Initialization code**: This is any code that appears at the top of the test script, before the setup code. It is run once per VU and is used to load files, import modules, configure the options used in the test, and for similar operations.

2. **Setup code**: The code runs once and is used to set up data that is shared by all the VUs that are running the tests. This code uses the following syntax:

    ```
    export function setup() { }
    ```

3. **VU code**: The code is run as many times as needed on each VU and is used to define the functions that will be run during a test. This code uses the following syntax:

    ```
    export default function (data) { }
    ```

4. **Teardown code**: The code is run once, but it will not run if the setup ends abnormally. It is used to process results and stop the test environments. This code uses the following syntax:

    ```
    export function teardown (data) { }
    ```

Now that we have a good understanding of using k6 to run tests, we need to consider the different ways we can install and run k6.

Installing and running k6

k6 is available in several package formats:

- Linux (.rpm and .deb)
- macOS
- Windows
- Containerized image
- Standalone binary for all platforms

Installation is very simple on all platforms, and full instructions can be found on the k6 website at https://k6.io/docs/get-started/installation/.

Running k6 is also very easy as all processes are triggered from the CLI. This is very well documented via the --help flag:

```
$ k6 --help
```

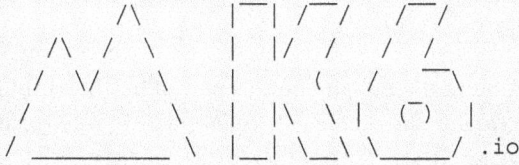

```
Usage:
  k6 [command]


Available Commands:
  archive     Create an archive
  cloud       Run a test on the cloud
  completion  Generate the autocompletion script for the specified
shell
  help        Help about any command
  inspect     Inspect a script or archive
  login       Authenticate with a service
  pause       Pause a running test
  resume      Resume a paused test
  run         Start a test
  scale       Scale a running test
  stats       Show test metrics
  status      Show test status
  version     Show application version
```

The k6 run and k6 cloud operations are used to run tests locally or via the k6 cloud, respectively. Here are some example commands using a test file called test.js:

- Run a single VU once:

  ```
  k6 run test.js
  ```

- Run 10 VUs with 20 iterations of the test being run across these VUs:

  ```
  k6 run -u 10 -i 20 test.js
  ```

- Ramp VUs from 0 to 50 over 20 secs, maintain the 50 VU count for 60 secs, then ramp down to 0 over 10 secs:

  ```
  k6 run -u 0 -s 20s:50 -s 60s:50 -s 10s:0 test.js
  ```

These commands could all have `k6 run` replaced with `k6 cloud` to use a k6 cloud runner instead of running the tests from the local machine.

Now that we've seen how to use k6 to perform load testing, let's wrap up.

Summary

In this chapter, we have explored two of the tools that Grafana offers as part of its observability platform: Pyroscope and k6. We learned how to search the profile data collected by Pyroscope and how to configure the client to collect that profile data. We also learned how to instrument applications, both using a native language SDK and using Lambda layers for serverless applications. Finally, we explored the new Pyroscope architecture and saw how it is very similar to Loki, Mimir, and Tempo. This new scalability should give Pyroscope the space to grow into a vital fourth telemetry type, making systems more observable.

With k6, we learned about various types of load or performance tests. We saw how we can easily write tests using the JavaScript language, using checks and thresholds to articulate vital measures for an application. We saw how to use `options` to manage how k6 runs its tests, and how to add the correct data and functions to our scripts to best make use of the test life cycle. Finally, we saw the process for installing and running k6, and how the simple operation even allows us to run the tool as part of a CI/CD pipeline to continuously load test applications to validate that their performance is meeting SLOs.

In the next chapter, we will bring together all of the tools, APIs, and knowledge to understand how to best support DevOps principles using Grafana.

14

Supporting DevOps Processes with Observability

This chapter will discuss the use of Grafana in two different aspects of the technology industry – **software delivery** and **platform operations**.

We will briefly introduce you to the **DevOps life cycle** as valuable foundational knowledge. Using this framework, we will guide you through the value of Grafana in each phase to enrich the software development process in your organization. We encourage you to spend time understanding where bottlenecks are in this process and focus resources on the most appropriate phase for your team or organization.

Platform operations are typified by using third-party applications. This removes about half of the DevOps life cycle, as those stages are conducted by a third party. You will be introduced to the considerations you should make for using Grafana during the deployment and operation of several types of platforms. We will look at collecting data from data collection tools in an observability platform and consider best practices around disaster planning for the failure of this business-critical system. We will look at the particular needs of the operators and users of platforms that provide **continuous integration (CI)** or **continuous delivery/deployment (CD)** capabilities to an organization, as monitoring these platforms can be challenging. We will discuss resources available to monitor databases, in-memory data stores, message buses, and web servers, covering how to install them efficiently and how these common tools have publicly available dashboards in Grafana to use. Finally, we will have a quick look at how this same pattern of monitoring platforms is applicable for some security tools.

This chapter will handle technical concepts but there are no requirements to have experience with individual tools, and the chapter should be accessible to anyone, regardless of background.

In this chapter, we're going to cover the following main topics:

- Introducing the DevOps life cycle
- Using Grafana for fast feedback during the development life cycle
- Using Grafana to monitor infrastructure and platforms

Introducing the DevOps life cycle

Before we explain what the DevOps life cycle is, let's consider the history of Agile, DevOps, DevSecOps, and platform engineering a little.

Iterative development practices were used as early as the late 1950s, but in the 1990s, several development methods were introduced as a reaction to development practices that were seen as heavyweight, micromanaged, highly regulated, and with a high risk of project failure. These new methods included **rapid application development (RAD)**, **Scrum**, **extreme programming**, and **feature-driven design (FDD)**. These all originated before the Agile Manifesto, but they are now known as agile practices. According to the Agile Manifesto, published in 2001, we prefer the following:

- Individuals and interactions over processes and tools
- Working software over comprehensive documentation
- Customer collaboration over contract negotiation
- Responding to change over following a plan

This indicates that, while there is value in the items on the right, we value the items on the left more.

Agile practices evolved from development practices, and they are focused on development teams, although there are a lot of crossovers with operational practices. These manifesto notions drove a lot of interest in practices such as test-driven development, CI, CD, and many others.

In the early 2000s, concerns around the separation of development practices and operational practices were highlighted (although these concerns were also raised through the 1980s and 1990s). These concerns coalesced in 2009 with the first *DevOps Days* conference. DevOps does not articulate a central philosophy such as Agile, but it suggests practices and measures that are intended to speed the delivery of working software to customers. A lot of these practices revolve around having developers, testers, and operators collaborate more closely, often by bringing them together in the same team. Similarly, development practices such as using version control systems (for example, Git) are adopted so operational concerns such as system configuration can become part of the shared understanding of a whole software system.

DevOps has several branches, extensions, and concepts. Here are some of them for those who are interested in reading further: **ArchOps**, **site reliability engineering (SRE)**, **DevSecOps**, **DataOps**, **12-factor apps** or **15-factor apps**, **infrastructure as code (IaC)**, **configuration as code (CaC)**, and **GitOps**.

A quote from Amazon CTO Werner Vogels back in 2006 became a bit of a rallying cry for the DevOps movement: *"You build it, you run it."* This has a lot of merit. Having the team who designed and wrote a product also be responsible for its operation should mean that incidents are resolved quicker and customer feedback can be heard and responded to. Teams can be more agile! When managed well and in the right organization, this can be a very effective way to operate. However, as the analysis by

Matthew Skelton and Manuel Pais in *Team Topologies* (`https://web.devopstopologies.com/index.html#anti-types`) shows, many anti-patterns can appear and lead to dysfunction in an organization. This approach can also lead to a significant cognitive load for development teams, which makes organizations less able to respond to change.

You might ask why we include this history when we are explaining what the DevOps life cycle is. The reason is to caution you that this life cycle is a tool and, in most organizations, a collection of processes; while they do have value, they should not be valued more than individuals and interactions. The way that teams tasked with managing a customer-facing software system will interact with an observability platform will differ significantly from a team tasked with managing the platform in support of the organization's goals. With this caution given, let's look at the DevOps life cycle as it gives us a good framework to discuss the many aspects of using an observability platform through the life cycle:

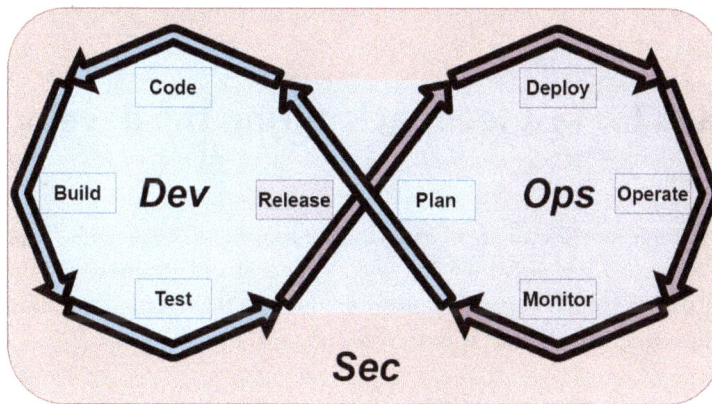

Figure 14.1 – The DevOps life cycle

> **Important note**
> There's isn't a clear definition of DevOps or DevSecOps. The DevOps life cyle itself covers development and operations while the security aspect wraps around all of that (and more), as shown in *Figure 14.1*.

Let's walk through each phase of this life cycle:

- **Code:** This is where new code is written in line with the specification given during the planning phase
- **Build:** This phase is where new code is built
- **Test:** New code is tested in various ways during this phase
- **Release:** The code is verified as ready to be deployed to production in this phase; any final checks or assurances will be performed here

- **Deploy**: The code is deployed to a production environment

- **Operate**: This phase is a continuous phase; the latest deployed release is run in a production environment

- **Monitor**: Any data collected from the release that is currently operating in production is gathered, as well as any feedback or user research, and is collated together to be used in the next planning phase

- **Plan**: During this phase, the team plans what future iterations of the product will contain

- **Security**: This is a continuous concern for the team in a DevSecOps approach and is the responsibility of all members of the team

Now that we have seen the DevOps life cycle, let's consider how we can use Grafana tools during each phase of this cycle.

Using Grafana for fast feedback during the development life cycle

In this section, we will consider how to use Grafana tools through each stage of the DevOps life cycle. Developing software can be risky and expensive, and observability platforms can also be expensive. Therefore, using the data from an observability platform to reduce the risks and expense of developing software is a great investment. We'll start with the *code* phase of the life cycle.

Code

To use Grafana in the DevOps life cycle, the system must produce useful data that can be used to understand the state of the system. To that end, the first act during the *code* phase of the life cycle is to *instrument* the system. Depending on the type of system we are working on, the method of producing data may look different:

- A **software application** would be instrumented by adding libraries or SDKs that produce data in a format agreed with the team(s) who collects this data. In some situations, this can even be achieved by the injection of instrumentation into the application, which can happen during the *deploy* stage of the life cycle. Organizations need to be clear on where this responsibility lies.

- A **cloud infrastructure** or **cloud platform** component would be instrumented by collecting data from the vendor.

- A **local infrastructure** or **local platform** component would be instrumented by collecting data in a format supported by the vendor of the component.

For a lot of systems, this may be all that is needed. However, there are times when an organization needs custom data from a system. Adding such instrumentation falls squarely in the *code* phase of the life cycle. However, when considering such activities, it is important to also ensure that the *plan* and *test* phases are considered. This can be achieved through activities such as agreeing on a data format and field definitions and implementing the code in a way that it can be tested in future iterations of the product (for example, domain-orientated observability).

The final area in which Grafana can help during the coding process is by being run directly against code as it is developed. Most, if not all developers, will run their code locally before it is committed to a version control repository. As Grafana is open source, it is very easy to implement a local development environment that produces and collects observability telemetry; we provided an example of this kind of environment when we explored live data in *Chapters 3, 4, 5,* and *6*. This wealth of information can feed directly back into the coding process as it happens.

The next phase of the life cycle is the *build* phase. We will skip over this as we deal with monitoring builds in a lot more detail when we talk about monitoring CI/CD platforms in the next section of this chapter. Let's talk about the *test* phase next.

Test

The *test* phase can cover a lot of different test types. While tests are typically managed by the CI/CD platform, such as the use of a testing framework or static analysis tools, the most common form of feedback in Grafana is to monitor the CI/CD platform itself. An additional approach for organizations or teams who want to track more information is to output time series data from the CI/CD platform into a **time series database** (**TSDB**). These kinds of custom approaches can often become like a complex Rube Goldberg machine, so we would caution you to be very mindful of what the value is to the organization, and we recommend that you research the market in case a more suitable product is available.

As the *test* phase moves into end-to-end tests, tools such as k6 really come into play (we discussed this in *Chapter 13*). Writing great repeatable tests for tools in this space can also offer a very valuable ability to run them during the *deploy* phase of the life cycle to confirm that the new code has been successfully deployed.

The *release* phase encompasses everything between completing testing and releasing code to customers. This often covers activities such as gaining approvals for the deployment from stakeholders or assurance teams. Let's have a look at how Grafana can help.

Release

Let's start discussing using Grafana for the *release* phase with a brief warning: many tools on the market may offer a better fit for organizations and teams, so we recommend that organizations do some research if they are having problems with their release processes.

Perhaps the biggest feature of Grafana that enables a smooth release step is the ability to show whether a new iteration of a product complies with the **service-level objectives** (**SLOs**) and **service-level agreements** (**SLAs**) for the product. Showing these metrics from a new iteration, especially when the product has been put under load by a tool such as k6, is a very powerful way to say that the new iteration behaves as expected.

The other feature that may be of interest to some teams is the ability to automatically build dashboards that contain HTML widgets. This can be used to automatically assemble a release dashboard with links to various artifacts such as test reports, tickets for included features, and similar.

The operational phases of the life cycle are the most associated with Grafana. Let's start looking at these with the *deploy* phase, in which code is deployed into a production environment ready for customers to access.

Deploy

The *deploy* phase will see a lot of changes occur, and the details of using Grafana will differ depending on how the system is deployed:

- Where an application is deployed to a Kubernetes cluster, Pods will be scheduled for termination, while new Pods using the newer version will be started. We might see Pods responsible for database updates scheduled, and various other aspects. When used as the repository for all telemetry from a Kubernetes cluster, Grafana can be used to visualize the deployment process in a way that suits the deployment team, from prebuilt dashboards to a custom dashboard specifically designed for a specific application deployment.

- Where applications are deployed directly to an operating system rather than a containerized environment, Grafana still offers detailed monitoring, with prebuilt dashboards for operating systems, common languages, web servers, databases, in-memory data stores, and many other tools.

These approaches provide **white box monitoring** of a deployment; a lot of organizations will also implement **black box monitoring** of the application during a deployment. Grafana can help here as well. By using **Grafana OnCall** to receive messages from an availability monitoring tool such as the Prometheus blackbox exporter, k6, or Pingdom, this stream of data can also be monitored during a deployment.

It is best practice to generate annotations when a deployment happens, which can be done via the API. Here is an example of an annotation added to a deployment of the OpenTelemetry Collector that caused an incident:

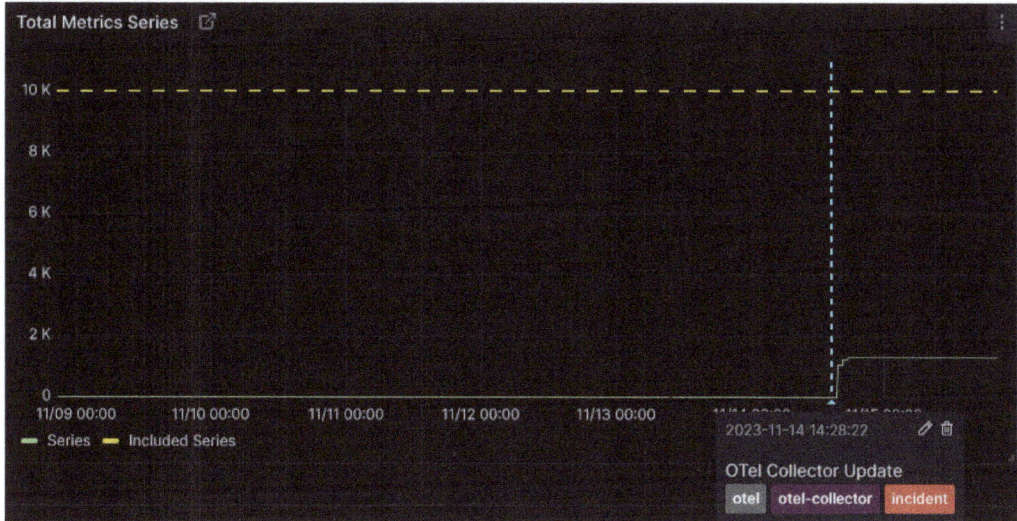

Figure 14.2 – Annotations in action

As the screenshot shows, Grafana will display contextual information about deployments on any visualization that has the option switched on. Annotations appear as a line on the chart and show information when hovered over; this contextual information can be tagged.

At their heart, CD platforms are code execution platforms, which means that any action that can be coded can be performed by a CD platform. We just talked about monitoring a deployment visually using a dashboard. This approach is great when deployments happen infrequently. When deployments occur much more frequently, it can be valuable to invest time in writing stages of a deployment where the state of the application being deployed is monitored. Loki, Mimir, and Tempo all offer query endpoints, which can be used to run queries as part of the scripted CD job. Effectively, this offloads the need to watch a dashboard to the CD pipeline, and rollback steps can be defined if the deployment fails. Some common examples of this use are as follows:

- Monitoring the error rate seen in the application logs.

- Checking whether login actions are successful. This would usually be tied to a smoke test to ensure that login events occur.

- Checking whether communication with downstream services is affected.

If these checks were to fail, the deployment could be quickly rolled back using automated procedures. This approach significantly reduces the **mean time to recovery (MTTR)** for such common issues and ensures that engineers can be focused on more valuable tasks during a deployment.

The gold standard for leveraging the tools provided by Grafana is to also deploy any updates to the dashboards used for a service with Terraform during the same deployment window as the code deployment. Adopting this practice allows for an easily repeatable process, moving from local development work through testing and into a production environment.

While exciting, the *deploy* phase is not the phase where code is in *normal* operation; that phase is the *operate* phase. Let's look at this phase next.

Operate

The *operate* phase is where the product is live in front of customers. The most important aspect of this phase is ensuring customers are getting a great service. This can be achieved by monitoring SLOs and SLAs, checking errors that may occur, responding to incidents, and helping customers in their use of the product. Grafana is primarily a tool that is used through the *operate* phase of the life cycle, so most tools in Grafana are targeted toward this phase. Some key components that will be used by all teams who use Grafana are **dashboards** and **alerts**. The ability to see how a user is interacting with a product is also very valuable to operational teams, such as customer experience or customer support teams.

We discussed in *Chapter 9* how **Grafana Alerting** and **Grafana Incident** can integrate with many systems. This capability is very helpful in creating a detailed incident response system – for example, by linking Grafana with ServiceNow so the creation, updating, and closing of incident tickets can be partially or fully automated, even with capabilities to collect chat communications to reduce the time needed to write up what happened during an incident for reporting.

We talked about using **Grafana Frontend Observability** in *Chapter 12*; when correctly implemented with distributed tracing, this tool allows customer-facing teams to reconstruct an individual user's session. This allows these teams to work quickly with the customer to understand the frontend problem they are experiencing and translate that into a trace path through the system to identify the source of the issue and get it to the right team quickly, with easy-to-digest information on what happened.

Let's consider how to use Grafana to monitor the system.

Monitor

Like the *operate* phase. the *monitor* phase is the phase in which using Grafana can really shine. The two biggest challenges are knowing what telemetry to use to answer a question about the product and whether the telemetry is being made available. While it would be impossible to list every potential question, here are some common questions, linked with the telemetry type that would be best suited to answer them:

- How are my customers interacting with my product?

    This is best answered by using real user monitoring, which we discussed in *Chapter 12*. This question could cover many similar questions such as what the uptake of a new feature is, and whether there are unvisited pages or features in the system.

- Are there particular functions that are slow?

  This can be answered by combining the timing information for requests from metrics with the detailed application information produced in logs. We discussed these in *Chapters 4* and *5*. For applications with downstream dependencies, this information can also be complemented with trace data, as discussed in *Chapter 6*.

- Why is a particular function slow?

  Often, this question will be answered through local testing, but this process may be significantly aided by using continuous profiling against a system with real or replayed requests. *Chapter 13* discussed continuous profiling in more detail.

- Is my application behaving as expected?

  This is best addressed by establishing clear **service-level indicators** (**SLIs**) and SLOs for the application; we outlined how to do this in *Chapter 9*.

- Is the service compliant with the SLOs/SLAs?

  This is typically answered by using metrics data. However, some indicators may be metrics derived from logs or tracing data – for example, creating a metric from logs of the number of errors seen.

- Is my infrastructure scaled correctly?

  This would be answered by collecting data from the infrastructure. How that is done may differ depending on the type of infrastructure:

  - For cloud infrastructure, this is done via an integration that provides logs, metrics, and sometimes tracing data

  - For on-premises infrastructure, the collection methods will vary

  We discussed this topic in more detail in *Chapter 7*.

- What is the long-term trend for something?

  The best telemetry type for long-term trending is metrics as they provide a default 13-month retention period. This means the best practice for such analysis is to produce a metric from the data you wish to track.

  Another approach would be to load data from Grafana into some form of data warehouse, but this is outside the scope of this book.

The real difference between the *operate* phase and the *monitor* phase is the aim of the use of Grafana. In the *operate* phase, the goal is to ensure that the system is functioning correctly for customers of the system. In the *monitor* phase, the goal is to understand and document how the system is functioning to feed into the *plan* phase to improve the system. Let's finish discussing the DevOps life cycle with the *plan* phase.

Plan

The *plan* phase takes input from many sources to help a team decide what the next priority for work is. The questions asked in the *monitor* phase, and any incidents or SLO breaches from the *operate* phase, are some of those sources. To help prioritize, it is common to consider things such as the following:

- How many customers are affected by a particular incident or potential improvement?

  The logs, metrics, and traces in Grafana can collect the data needed to answer this. This is true even for changes that have been sourced from other places such as user feedback.

- How close to capacity is a component of the system, or how much time is there to address a bottleneck before it begins creating incidents or performance degradation?

Identifying bottlenecks before they become critical can be done by using k6 to load test the system with spike testing, stress testing, or even testing it to breakpoint.

The DevOps life cycle is very focused on teams who are developing software. It's common for organizations to use software provided by third parties to provide internal platforms. There is a lot of crossover with the *deploy*, *operate*, and *monitor* phases, but let's take a more detailed look at using observability with some of these platforms.

Using Grafana to monitor infrastructure and platforms

Teams who work with third-party infrastructure and platforms are well supported by the tools from Grafana and OpenTelemetry. We'll consider a few major types of platforms; observability, CI, CD, infrastructure and resource, and finally, security platforms. The *deploy*, *operate*, *monitor*, and *plan* phases should all be understood for these platforms and the points made in the previous section for these phases are relevant to these kinds of platform products. Let's start by considering observability platforms.

Observability platforms

Teams who manage observability platforms have a responsibility to offer a platform that demonstrates best practices by having well-documented SLIs and SLOs, easy-to-find dashboards, and a dependable incident management process.

Helpfully, there are dashboards available through the Grafana Dashboards community portal that provide very detailed views of the OpenTelemetry Collector and the data flows as they pass through the Collector. Deciding what aspects of the Collector are most important to your organization and publishing them is a step that should be taken by any team that manages observability collection.

An important consideration for managing an observability platform is the disaster management process for the loss of the platform. While this scenario is unlikely, it is much better to have a tested plan than to try to come up with one when the platform is on fire – this is advised after a very painful experience. Usually, such a disaster plan can be simple – for instance, the ability to create a Prometheus instance or even a full Grafana stack in each cluster will give organizations the capability to continue operating in the event of the **software-as-a-service** (**SaaS**) platform they use being down.

A related plan that should exist is how noisy data sources are controlled. Compartmentalization of production data from other sources is a best practice. Sometimes, the financial or capacity cost of a noisy data source could be a business interruption. These risks can be managed in several ways, such as revoking API keys, adding filtering to collectors, or even more extreme measures such as switching off the data source.

Let's consider CI platforms next.

CI platforms

CI platforms cover a lot of different tools, such as Github Actions, GitLab CI/CD, Jenkins, Azure DevOps, Google Cloud Build, and similar. We believe the most common question asked of CI platforms is "*Why did my build fail?*". Giving engineers tools to debug their builds is very important for such a platform. Often, this feedback can be seen in the CI platform itself. However, for some types of failures, it may not be obvious, such as a runner that failed, a noisy neighbor, or some other issue. In these cases, having data collected from the CI platform itself can be very useful.

Due to the nature of the CI platform, data collection usually needs to be tailored to the platform:

- Platforms provided by cloud vendors would usually be instrumented by collecting the logs and metrics from the platform in the vendor's own tooling (for example, AWS CloudWatch, GCP Operations Suite, or Azure Monitor) and then sending them on to a Grafana instance if appropriate.

- Other platforms will probably need to have an agent installed. We discussed this process in *Chapter 13*. For reference, the OpenTelemetry Collector is provided via a Docker image, Alpine image (APK), Debian image (`.deb`), Enterprise Linux image (`.rpm`), and as a general image (`.tar.gz`), which includes executables for macOS (Intel and ARM) and Windows. The Grafana Agent is provided as a Docker image, Debian image (`.deb`), Enterprise Linux image (`.rpm`), SUSE image, macOS image (via Homebrew for Intel and ARM), and Windows Installer (`.exe`).

Once an agent is installed, the configuration should be managed to give the best support for the nature of the integration work that is carried out on the platform. We recommend using one of the automation tools discussed in *Chapter 10* to manage this.

Logs and metrics are the prime data components to capture, as CI platforms do not typically need distributed tracing. One thing to consider as a team adds observability to a CI platform is whether the leadership team wishes to track higher-level business metrics – for example, lead time for changes, defect counts, or similar. For those of you who want to look further into these ideas, we recommend looking at Google's **DevOps Research and Assessment** (**DORA**) team reports (`https://cloud.google.com/devops/state-of-devops/`). These kinds of considerations would usually need to be agreed upon across several teams, so having a clearly documented definition of how they are calculated and collected is vital. This kind of data collection may or may not be done in the observability tooling. It is best practice to separate the data from CI platforms from business-critical workloads. This can easily be achieved by dedicating a separate Grafana stack for CI workloads. There are publicly available dashboards for these systems as well.

Now that we've seen how to monitor a CI platform, let's consider the deployment platform.

CD platforms

CD platforms often have crossover with CI platforms; we're considering them separate as they are different aspects of the overall system. These platforms use tools such as Jenkins, GitLab CI/CD, AWS CodeDeploy, ArgoCD, FluxCD, and similar. For infrastructure deployment, they may also include tools such as Terraform Cloud, Atlantis, or Spacelift. There are two main groups of CD tools: **push systems** and **pull systems**. We'll discuss them separately in this way as the data collection processes differ. With either deployment method, a very important aspect of integrating with Grafana well is to record an annotation in Grafana. We discussed this in more detail when we talked about the *deploy* phase of the DevOps life cycle, but this contextual information can save huge amounts of time during troubleshooting, and ultimately, provide a better customer experience.

Pull systems in Kubernetes also use the term *GitOps*; such systems typically use tools such as **ArgoCD** or **FluxCD**. As these tools are deployed into an existing Kubernetes cluster, the observability stance is very simple, in that the service will have data collected by the existing collection infrastructure in the cluster. ArgoCD provides metrics in Prometheus format, and there are several dashboards publicly available. It's also possible to extend the data collection via other tools in the Argo group of tools. FluxCD provides Prometheus metrics that can be extended with kube-state-metrics as well. The tool also provides logs and produces Kubernetes events as well. There are other pull systems outside of Kubernetes, such as **Chef** and **ansible-pull**, but due to the low prevalence of these tools, we'll not discuss them here.

Push-based CD platforms have one or more central systems that connect to the deployment target and run the deployment process. Jenkins is perhaps the classic example here, but systems such as GitHub Actions and GitLab CI/CD also fall into this category. You may notice that these tools were also mentioned previously when we considered the CI platform. Unsurprisingly, these tools are monitored in the same way, whether they are used for integration tasks or delivery/deployment tasks. When the use of these tools has a mix of integration and delivery/deployment tasks, monitoring the actions of the platform is very important from a security perspective as these systems will often consume third-party libraries during the integration phase, which opens the system to supply chain attacks. Combining such an attack with high-level access to production on a single system is a very real threat to organizations.

We've now considered how to monitor the platforms that build and deploy software. Let's consider a wider group of systems next. We'll consider data storage and message queue systems in this next section.

Resource platforms

We're using the term *resource platform* to describe the types of backend systems that an application may depend on. These might include databases, in-memory data stores, message buses, web servers, or similar. These platforms are an odd case, as the responsibility for the system can reside in many different areas of the organization. Commonly, either a software delivery team would be responsible, or sometimes a centralized team would be responsible. We will attempt to ignore this complexity by talking in general terms about how to ensure these tools are observable.

There are a few places to start looking at monitoring these systems:

- **OpenTelemetry Collector contributed receivers**: There are contributed receiver modules for a vast array of resource platforms. Deploying these receivers is very simple:

  - In Kubernetes, the collector can be deployed as a sidecar to the service, or as a dedicated agent in the cluster or namespace that forwards telemetry on to a gateway. An example of a dedicated agent is shown in *Figure 11.5* in *Chapter 11* where the agent is used to collect metrics from the Kubernetes cluster.

  - In a virtual or bare metal installation, a dedicated OpenTelemetry Collector needs to be deployed, although this can often be done on the instance being monitored with no performance degradation.

- **Prometheus modules**: These are modules that allow Prometheus to scrape data from a lot of resource platforms. These can simply be deployed to a Prometheus instance and configured to connect to the platform that needs to be monitored.

Once metrics are collected from these systems, Grafana offers a wide range of prebuilt public dashboards for them. The wide availability of these dashboards means the time to value is very good.

One thing to highlight to prevent confusion with these systems is that this type of data collection is not the same as using the system as a data source in Grafana. A lot of these systems, especially databases such as MySQL, PostgreSQL, and MongoDB, can be used as a Grafana data source. A data source connects to the system and allows users to query the data in the system. The tools we are discussing here connect to the system and query operational metrics from it. These metrics can then be used to provide SLOs and transparency of the operation of the system to other teams.

Security platforms

We will not delve too deeply into **security platforms** as they could fill an entire book on their own. However, it's worth noting that several tools, such as **Falco**, **Open Policy Agent**, **kube-bench**, **Trivy**, and others, have methods of exposing metrics related to their operation, which can be consumed by Grafana in some way.

There is also a very big crossover of concerns of observability platforms with cyber security platforms. Both platforms consume log data, which can lead to the running of multiple agents to collect this data. A more cost-effective solution could be for these teams to work together on a shared pipeline of this data that supports both teams' operations. Such a pipeline should be monitored closely as it could present a significant risk to the organization.

We've now considered using Grafana both in the software application life cycle and for the management of infrastructure and platforms.

Summary

In this chapter, we have considered how you can use Grafana through the DevOps life cycle. You learned about deploying Grafana in a local environment to speed up development time by getting instant feedback on the performance of the code. We looked at the testing phase, and you learned how using tools such as k6 can provide great repeatable tests that can even be used as an application is deployed. During the release phase, Grafana can be used to demonstrate various aspects of an application to the stakeholders who approve your releases. We saw how deployments can have their risks reduced by leveraging SLOs and black box monitoring. We also saw how using Grafana annotations can improve the visibility of deployments occurring. The *operate* and *monitor* phases use Grafana in very similar ways, which have been covered in this book. You were introduced to the difference in the aim of these two phases, with the *operate* phase being concerned with the correct functioning and the *monitor* phase being concerned with how to improve the customer experience. Finally, we talked about how Grafana can be used to have a data-driven discussion during the planning phase of a software tool.

We then considered how Grafana can also be used with various types of platforms. We introduced you to using Grafana to monitor your observability platform, effectively demonstrating the principle of using your own product or "*eating your own dog food*," and acting as an example of best practice to an organization. You saw how to use Grafana with your CI/CD platforms, so engineers in an organization have a lot of data to understand how their builds and deployments are working. We then discussed how to get operational data from many systems used across the industry, such as databases, in-memory data stores, message buses, and web servers. You learned that the best approach is to look for available data collection tooling and publicly available dashboards. The final kind of platform we looked at was a security platform, where you saw that some tools also surface data in Prometheus or OpenTelemetry format, which can be consumed by Grafana. Where this is available, prebuilt Grafana dashboards are also available, which significantly reduces the time to value for using these tools.

We have nearly reached the end of the book. The next chapter will cover best practices and troubleshooting techniques. You will look at some specific items around data collection and the Grafana stack as well as general guidance on common pitfalls in observability. We will also discuss interesting future trends.

15

Troubleshooting, Implementing Best Practices, and More with Grafana

We are nearing the end of our journey in *Observability with Grafana*, so this is a good time to revisit some of the things we have learned. In this chapter, we will review the best practices for data collection and look at troubleshooting techniques that can help get your telemetry into Grafana. We will then move on to the Grafana stack for more best practices and troubleshooting with your telemetry backend, alerts, and dashboards. We will identify some of the pitfalls of observability and how to avoid them. Then, before wrapping up, we will look to the future and explore potential trends on the horizon.

In this chapter, we are going to cover the following main topics:

- Best practices and troubleshooting for data collection
- Best practices and troubleshooting for the Grafana stack
- Avoiding pitfalls of observability
- Future trends in application monitoring

Best practices and troubleshooting for data collection

Throughout this book, we have repeatedly talked about the importance of preparation, setting objectives, and defining requirements. Doing the upfront thought work cannot be underestimated, but sadly, it is often completely overlooked. There are far too many occasions where I have investigated a company's observability platform and discovered collection agents have been deployed without thought, opening the floodgates to send everything into the backend. The price is an excess of undefined telemetry, making it hard to do your job and causing expensive operating and storage costs.

In this section, we'll be sharing some best practices for data collection, together with some useful troubleshooting tips. Let's first look at preparation activities for data collection.

Preparing for data collection

Observability starts with a need or desire to monitor and observe systems, and to do that, we need data. However, too much data (and data for the sake of data) can break your system's observability, which is why it is important to prepare. Set your objectives, considering what you want from your platform and who it is for. Instead of planning for the *what if*, plan for *what is*. You can always add to it later, which is much easier than taking away after the fact.

In *Chapter 1*, we introduced observability personas. You can use these to gather platform requirements. Ideally, you want to be able to answer the following questions:

- Who are my observability customers?
- What log formats do I want or have in my system?
- What is the origin of the data?
- What metrics are important for our **service-level objectives (SLOs)**?
- Will tracing instrumentation help solve observability problems?
- What data do I need for dashboards?

These answers will help you determine which agent technology works best for your use case, what log formats to use, and other important decisions.

Let's look at some of the decisions we need to make for our data collection.

Data collection decisions

Several critical decisions must be made early on that can impact your entire observability journey. It is important to take the time upfront to make these decisions, as it becomes a lot harder and more expensive to change them later. In most scenarios, your choices are reduced, and you have to work with existing data. That does not stop you from identifying and communicating a new standard to aim for. Standardizing data collection across the entire organization provides a framework that supports engineers in compliance. Here, we group some of those factors together to help you process them:

- **Logs**:
  - Choose a log format that can be extended so you can deliver quickly and enhance later.
  - Select labels carefully, considering Loki's performance and cardinality. You can always extract additional fields as labels during querying.
  - Consider whether valuable metrics can be created from logs (to maximize the value).

- **Metrics**:

  - Identify important metrics, and drop what you don't need. This can help with metric cardinality, too. If you cannot drop the whole metric, just dropping some of the highest cardinality labels can help a lot.

  - Choose the protocols that provide the data you need (remember, there are variations, so read the documentation for each carefully).

  - If using verbose metric protocols, ensure protection is in place (e.g., histogram bucketing) to restrict the ability to flood your system.

  - Add context so you can correlate metrics with traces.

- **Traces**:

  - Ensure that the accuracy of spans and traces is implemented and validated

  - Balance the performance and cost impact with a mitigation strategy (sampling, filtering, and retention)

- **Instrumentation libraries**:

  - Research them well. If you are using a library, you want it to be maintained and supported going forward.

- **Telemetry collector**:

  - Run proofs of concept to validate what works with your technology. You don't want to fall foul of permission constraints restricting the choice of collector on your route to production.

  - Consider the support model that comes with the collector technology, if any.

  - What are your business needs from a collector?

Now that all the decisions have been made, let's look at what to do if your telemetry is not showing up.

Debugging collector

Within the observability platform, it can be difficult to isolate problems with your telemetry getting to the backend, especially if it is sent through different connections and components in your telemetry pipeline.

Here are a few steps that can help diagnose issues:

- Check for error messages in the collector logs

- Look for data rejections by Grafana, such as the following:

  - Sample too old

  - Trace too large

  - Ingest rate limiting

- Validate authentication credentials (token expiry and permissions)

- Verify how telemetry is being ingested (ports and protocols)

- Analyze whether the telemetry is modified with sampling or redaction

- Identify which exporters are being used to send telemetry to the next stage

- Validate the telemetry format

- Identify the next hop and validate its configuration

- Test simple network connectivity and identify whether firewall rules or network policies are restricting data flow

We should, hopefully, have data in the backend now, so let's look at some best practices with Grafana.

Best practices and troubleshooting for the Grafana stack

As with the previous section, the importance of preparation applies to the Grafana stack, as it does with any good system design. You are making decisions that affect your users, your data, and your costs. In this section, we'll be sharing some best practices for preparing Grafana, together with some useful tips for debugging. Let's first look at preparation activities for your Grafana stack.

Preparing the Grafana stack

When designing your platform, consider what you are sending there and who your users are. With your stack, it is important to consider wider subjects such as corporate authentication integration, as well as best endeavors to right-size your stack and implement processes to monitor that usage.

Ideally, you want to be able to answer the following questions:

- Who needs to use the platform?

- Where are they using the platform?

- What data retention requirements exist (if any)?

Grafana stack decisions

There are important decisions to be made before you fully embrace your Grafana stack. Not all of them are related to observability; some are affected by regional governance and some are company policy. Working on and prioritizing these decisions earlier will help with the smooth running of your platform.

Here, we group some of those factors together to help you process them:

- **Architecture**:

  - Does your use case require single or multiple stacks – for instance, in the case of data residency or the separation of development and production systems?

  - Is any data restricted and requires specific permissions, such as **Personally Identifiable Information (PII)** and the **General Data Protection Regulation (GDPR)**?

  - Are there global latency issues to consider?

  - Are there any user audit requirements?

- **Management**:

  - Can the backend be outsourced to ease administration efforts?

  - Do you have the ability to fix problems in a timely manner in the case of an incident?

  - Would IaC provide platform control?

- **Authentication**:

  - Can the platform be integrated with an authentication provider to make user management easier?

  - Do you have team-specific permissions that might cause issues with an identity provider?

- **Data retention**:

  - You need to create data retention policies to understand why telemetry data is needed and for how long

  - Metrics are usually cheaper to store – you can generate these from logs with recording rules rather than store more expensive logs for longer

Debugging Grafana

There are a few areas or touchpoints where problems can occur with your Grafana backend. The following techniques should help you get back on track or at least provide a point of reference to help you debug further:

- **Data collection**: If you are having difficulties getting telemetry into Grafana, drop back to the APIs to send test data to validate network routes, API keys, and so on. This can be done with tools such as Postman or quite simply on the command line with `curl`.

- **Data querying**: Sometimes, queries do not give you the results you expected or provide them in a timely manner. Here are a few steps that can help solve problems or diagnose them:

  - **Break down large queries**: Start small and build up your results, validating at each step by using short time ranges and stronger filters

  - **Check function order**: The order of query functions affects the outcome, in the same way that the order of equations affects the result in math

  - **Speed up slow results**: Results can be obtained more quickly by following these practices:

    - Check how many data points there are

    - Limit the maximum data points returned in query options

    - Increase the minimum interval timeout in query options

    - Use `group by` functions to reduce the result set

  If you are trying out different variations of your query, the **Query history** button will remember them for you.

- **Problem discovery**: It is not always obvious where problems lie – unless, of course, people report them. Grafana provides a set of dashboards to support your success with Grafana Cloud. Here is a screenshot of the Grafana Cloud **Dashboards** list:

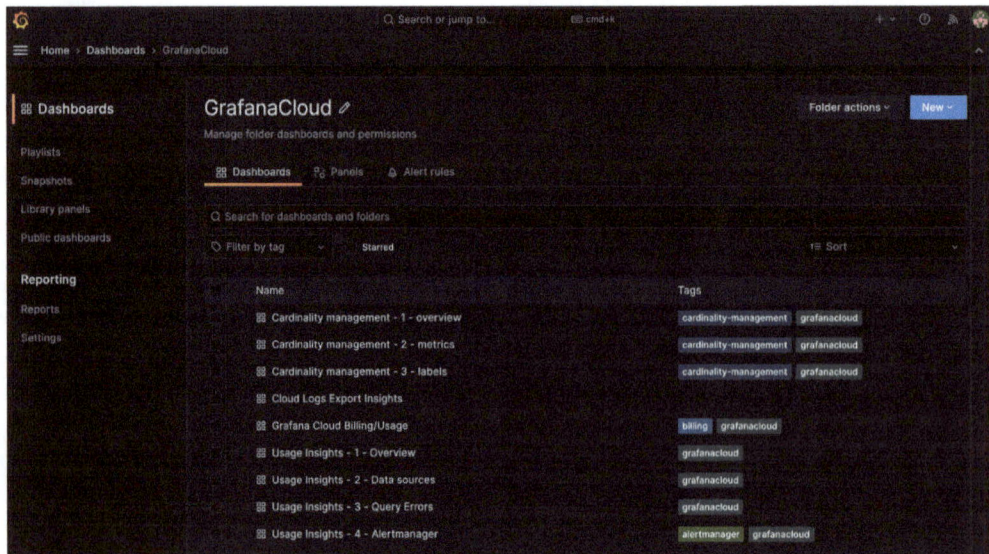

Figure 15.1 – Grafana Cloud Dashboards

When diagnosing problems with your Grafana data and dashboards, the **Usage Insights** dashboards can help. Navigate to the **Usage Insights – 1 – Overview** dashboard and scroll down; you will find three panels:

- **Top 10 dashboards with errors**: This lets you know which dashboards are encountering errors of some form.

- **Top 10 data sources with errors**: This reports the Grafana data sources that have issues. This is useful to diagnose errors with queries, or in communicating to the backend data source.

- **Top 10 users seeing errors**: This identifies your platform users who are encountering problems inside Grafana. This is helpful when investigating platform stability.

These panels guide you to the issues being encountered with drill-down links into other dashboards to show specific errors to aid diagnosis.

In particular, the **Grafana Cloud Billing/Usage** dashboard is very important in showing why samples might be rejected due to hitting account limits. Additionally, the **Cardinality management** dashboards can help highlight problems where label values are causing issues.

Let's now look at problems that we could encounter in the wider subject of observability.

Avoiding pitfalls of observability

We have identified several areas throughout this book where you need to sit down and think about your approach. All of these considerations contribute to your ongoing success with your observability platform. We will now call out a few of these, with some guidelines to follow:

- Treat your platform as an evolving development; start with the basics and build upon that, continuously reviewing where you are. Reducing the time to value will ensure return on investment is being realized.

- Collect requirements from multiple customers and then plan a roadmap that will deliver more value to more users to ensure the adoption and support of your platform.

- Monitor costs, paying close attention to the data collection stage, where it is cheaper to fix. Use your observability tools to help.

- Fix cardinality early, and develop standards that teams can work to that will control the problem. Additionally, you can apply governance that restricts data collection into the platform if it does not adhere to standards.

- Isolate high-load, low-value environments (especially performance test environments) to protect business-critical system observability. You can build cheaper, short-lived observability systems for these environments to keep costs under control.

- Define a 6- and 12-month roadmap. This will help you plan for and adapt to industry changes. Observability is moving fast, with new developments coming all the time; being agile will help you navigate them.

- Monitor platform usage. Grafana provides some excellent dashboards, as shown in *Figure 15.1*, that help you understand what is being used and how. Additionally, you can enhance this by improving the monitoring of your collector estate to create visibility of your entire telemetry pipeline.

Let's now look to the future and think about where observability could be heading.

Future trends in application monitoring

It is difficult to present a perspective on future trends without being opinionated, so the following opinions reflect my beliefs on where the industry is heading:

- **Cost reduction**: At the time of writing, companies are actively looking at ways to reduce operational expenditure. There is a lot of scope for cost reduction in observability systems. Grafana is leading the charge with this. Take a look at the following:

  - The **Grafana Cloud Billing/Usage** dashboard to get an understanding of Grafana Cloud spend. This dashboard is part of the list shown in *Figure 15.1*.

  - **Adaptative Metrics** (`https://grafana.com/docs/grafana-cloud/cost-management-and-billing/reduce-costs/metrics-costs/control-metrics-usage-via-adaptive-metrics/`) for solutions to reducing costs.

  - **Log Volume Explorer** (`https://grafana.com/docs/grafana-cloud/cost-management-and-billing/analyze-costs/logs-costs/analyze-log-ingestion-log-volume-explorer/`) to discover sources with excessive log collection.

- **Artificial intelligence** (**AI**): AI has matured recently to a point where it will soon become a key part of observability platforms. Grafana recently released generative AI features for the **Dashboard** panel title and description text and incident response auto summaries. You can see and read more here: `https://grafana.com/blog/2023/08/28/generative-ai-at-grafana-labs-whats-new-whats-next-and-our-vision-for-the-open-source-community/`. Hot off the press as we go to press with this book, Grafana announced the acquisition of **Asserts.ai** delivering root cause analysis solutions into Grafana; you can learn more here: `https://grafana.com/blog/2023/11/14/grafana-labs-acquires-asserts/`.

- **Tool enhancements**: Capabilities to improve the relationship between developers and observability, easing the adoption earlier in the development life cycle. Grafana recently released live dashboard development with a VS Code Grafana extension: `https://marketplace.visualstudio.com/items?itemName=Grafana.grafana-vscode`.

- **OpenTelemetry standard maturity**: Increased third-party development supporting OpenTelemetry as standards are agreed upon. Vendor-neutral solutions are becoming more popular, helping reduce overall costs (operational and management).

- **Collector management**: Here are some configuration and control tools for agent technology:

  - **Open Agent Management Protocol** (`https://opentelemetry.io/docs/specs/opamp/`) for the remote management of large fleets of data collection agents. No more re-deployments to block a metric or add a new receiver.

  - **Grafana Agent Flow** brings new ways to build complex telemetry pipelines for the Grafana Agent. It includes a visualization of the Agent pipeline, which is perfect for understanding complex configurations.

- **Platform engineering**: Advancements in platform engineering will help increase observability adoption and development as dependency on telemetry increases.

We will now wrap up this chapter and the book. You should now have the knowledge to implement, troubleshoot, and manage Grafana for your observability journey.

Summary

In this chapter, we have gone over best practices and troubleshooting techniques for data collection and the Grafana stack. We have looked at ways to avoid the pitfalls of observability as a whole, wrapping the chapter up with a look at future trends in application monitoring. These sections should have given you insights that will support your observability platform being a success, and help you get value from your platform for your users quicker and more efficiently. Now that we've completed the final chapter in our journey together through *Observability with Grafana*, let's take a moment to review our key learnings.

In the first part of the book, we introduced observability and Grafana, along with a look at application and infrastructure instrumentation. We closed that part by setting up the OpenTelemetry demonstration application in your own Kubernetes environment.

In part two, we worked through different telemetry types that you will encounter with Grafana – Loki for logs, Prometheus (Mimir) for metrics, and Tempo for traces – together making **LGTM (Loki, Grafana, Tempo, Mimir)**, which you will see all over the Grafana website. We then looked at integrations with Kubernetes (which we used with the demo app throughout the book), AWS, GCP, and Azure.

In part three, we worked with Grafana more, presenting data with dashboards and building an incident management process with alerts. We then explored IaC for configuring Grafana, followed by a look at the architecture of the Grafana stack.

In our final part, we talked about real user monitoring with Grafana frontend observability, application profiling with Grafana Pyroscope, and performance testing with Grafana K6. We closed the book with a look at how DevOps can be supported with observability, followed by some best practices and troubleshooting in this chapter.

Technology in the observability space, and especially with Grafana, moves fast. Hopefully, we have provided you with some timeless approaches and techniques that you can develop to support your observability work. Your new friends, the observability personas, will be there to lend a hand when you need it. Thank you for allowing us to be part of this journey with you. Good luck!

Index

A

access levels 236
   setting 256-258
access policies 54
access tokens 54
Adaptive Metrics
   reference link 312
advanced dashboard techniques 186
   layout tricks 186, 187
   tech tips 187, 188
agent-operator chart 235
agile practices 292
Agua Clara locks on Panama
      Canal, case study 5, 6
   distributed traces 8
   logs 7, 8
   metrics 7
   telemetry types 9, 10
AIOps 201
alert groups 214
alerting rules 21
alerts 298
   managing, with Ansible 240, 241
   managing, with Terraform 240, 241
alternatives, to Grafana stack 22
   data collection 23
   data storage, processing and
      visualization 23, 24

Amazon CloudWatch 155
Amazon CloudWatch dashboards
   using 157
Amazon CloudWatch data source 155
   configuring 155, 156
Amazon CloudWatch query editor
   using 156
Amazon Web Services (AWS) 147
Ansible 225, 228
   used, for managing alerts 240, 241
   used, for managing dashboards 240, 241
   using, for Grafana Cloud 237
ansible-pull 302
application monitoring
   future trends 312, 313
application performance
      monitoring (APM) 169
ArgoCD 302
artificial intelligence (AI) 312
Asserts.ai 312
   reference link 312
automatic instrumentation 43, 44
automation
   architecting 254
automation tools 3
AWS Distro for OpenTelemetry
   reference link 139

AWS Identity and Access
  Management (IAM) 155
AWS integration 155, 157
  configuring 157, 158
AWS Secrets Manager 238
AWS telemetry
  visualizing, with Grafana Cloud 155
Azure DevOps 21
Azure Monitor dashboards
  using 170
Azure Monitor data source 166
  configuring 166
Azure Monitor query editor
  metric queries 168
  reference link 168
  using 167-169
Azure Resource Graph (ARG) 169

B

backend data
  used, for pivoting frontend data 271, 272
baggage 140, 143
bare metal/compute 46
  telemetry examples 46
Basic Search mode 133
black-box monitoring 296
  RUM 204
  synthetic monitoring 204
black-box monitoring techniques 203
blameless postmortem 208
built-in aggregation operators 98
  examples 99

C

cardinality 39, 107
CD platforms 302

ChatOps 219
Chat with Support 54
checks
  used, for writing test 286
Chef 302
child operator or > 137
CI platforms 301
Cloud Advanced 56
Cloud Free subscription 53
cloud infrastructure/cloud platform 294
Cloud Pro 56
CloudWatch metrics 159, 160
cluster agent 251
collection infrastructure
  automating, with Ansible 228
  automating, with Helm 228
command-line interface (CLI) 285
Common Event Format (CEF) 31
communication channels 201
Community Forums 54
Community.Grafana
  reference link 238
compactor 145
comparison operators, PromQL 110-114
complex system 3
complicated system 3
conditions 217
consumers
  telemetry, sending to 258, 259
containerization tools 254
container orchestration tools 62
  installing 62
context propagation 140
continuous delivery/deployment (CD) 291
continuous integration (CI) 291
continuous profiling 275, 276
  Pyroscope, using for 276
control theory 4

core Grafana stack 19
  Grafana 20
  Grafana Agent 20
  Loki 20
  Mimir 19
  Tempo 20
Core Web Vitals 204, 268
  Cumulative Layout Shift (CLS) 204
  First Input Delay (FID) 204
  Largest Contentful Paint (LCP) 204
  reference link 269
cost of goods sold (COGS) 244
cost optimization 147
counter metric 35
  examples 35
Cross Origin Resource Sharing (CORS) 266
custom configuration 273, 274
customer communication 207
Custom Resource Definitions (CRDs) 233

D

dashboards 175, 298
  case study 191-195
  creating 176-180
  developing 180-182
  folders 190
  managing 189-191
  managing, with Ansible 240, 241
  managing, with Terraform 240, 241
  objectives, identifying 185
  organizing 189-191
  permissions 191
  tags 190
  users and needs 185, 186
data architecture
  defining 244-246
Databricks 21

data collection 248-252
  preparing 306
data collection and metric protocols
  DogStatsD 120
  OpenTelemetry Protocol (OTLP) 120
  Prometheus 121
  Simple Network Management
    Protocol (SNMP) 121
  StatsD 120
data collection decisions 306
data collection decisions, factors
  instrumentation libraries 307
  logs 306
  metrics 307
  telemetry collector 307
  traces 307
data collection systems 227, 253
data production 247, 248, 253
data production systems 226
data production, testing methods
  demo applications, using 255
  pre-recorded datasets, using 255
data production tools 255, 256
data source 68
data storage 252
data storage architectures 121
  Graphite architecture 122
  Mimir architecture 123, 124
  Prometheus architecture 122, 123
data storage systems 227, 253
  features 227
data types, PromQL
  instant vectors 108
  range vectors 109
  scalars 110
data visualization 252
data visualization systems 227, 253
  features 227

decolorize 90

descendent operator or >> 137

DevOps Anti-Types
   reference link 292

DevOps life cycle 291, 292
   phases 293, 294

DevOps Research and Assessment
   (DORA) 301

Diego Developer 13
   goals 13
   interactions 13
   needs 13
   pain points 14

distributed traces 8

distributed tracing 40, 140

distributed tracing, best practices 42
   accuracy 43
   costs 43
   performance 42

distributed tracing protocols 41
   features 41

Docker 62

Docker Desktop
   installing 62
   installing, on Linux 63
   installing, on macOS 62
   installing, on Windows 62

DogStatsD 119, 120

Domain-Oriented Observability
   reference link 246

drop labels expression 95

E

escalation chains 218

events 10

exemplars 125
   using, in Grafana 125, 127

Explorer query types
   metrics query editor 164, 165
   SLO query builder 165

Extended Berkeley Packet Filter
   (eBPF) client 279, 280

extreme programming 292

F

Faro 22

Faro Web SDK 263

feature-driven design (FDD) 292

Fiddler 255

Filelog Receiver 151

flame graphs 278

FluxCD 302

Four Golden Signals
   reference link 193

fractal concept 208

frontend data
   pivoting, to backend data 271, 272

Frontend Observability enhancements 274
   custom errors 274
   custom logs 274
   custom measurements 274
   frontend tracing 274

functions 217, 218

functions, PromQL
   reference link 114

G

gateway agent 251

gateway architectures 250

gateway service 250

gauge metric 35
   examples 35

General Data Protection
    Regulation (GDPR) 309
generative AI at Grafana Labs
    reference link 312
GHZ 255
GitLab 21
Golang RE2 syntax 88
Golden signals 203
Google Cloud Monitoring 162
    data source, monitoring 162, 163
Google Cloud Monitoring dashboards 165
Google Cloud Monitoring query
    editor 163, 164
Google Cloud Platform (GCP) 147
Google Remote Procedure Call
    (gRPC) 120, 204
Grafana 19, 20
    automating, benefits 226
    basic roles 257
    custom roles 257
    Enterprise plugins 21
    exemplars, using 125, 127
    fixed role definitions 257
    Incident Response and
        Management (IRM) 21
    used, for monitoring Kubernetes 148
    using, to monitor infrastructure 300
    using, to monitor platforms 300
    visualizations, using 183-185
Grafana Agent 19, 20, 228
    and Ansible 235
    installation, automating 235
grafana-agent chart 235
Grafana Agent Helm charts 235
Grafana Alerting 54, 197, 209, 210, 298
    alert rules 210-213
    contact points 213
    groups and admin 213

notification policies 213
silences 213
Grafana Ansible collection 238
Grafana API
    exploring 239, 240
    used, for obtaining grips 236
Grafana Cloud 50
    account setup 50-53
    Ansible, using for 237
    AWS telemetry, visualizing with 155
    Terraform, using for 237
Grafana Cloud API
    exploring 236
    functions 236
    reference link 237
Grafana Cloud Frontend Observability 263
Grafana Cloud Portal 53
    Billing 55, 56
    Org Settings 56
    Security 54
    Support 54
Grafana Cloud Profiles 276
Grafana Cloud RBAC 256
Grafana Frontend Observability 298
    setting up 265-269
grafana.grafana collection
    reference link 235
Grafana Incident 21, 197, 222, 298
Grafana instance 56, 256
    Administration panel 60
    Alerts & IRM section 58
    Connections section 59
    Dashboards section 57
    Explore section 58
    Observability section 59
    Performance testing section 59
    Starred dashboards section 57

Grafana Labs 19, 21, 24

Grafana Loki 81, 82
  architecture, exploring 99-101
  core components 100, 101
  functionality 100, 101

Grafana OnCall 21, 197, 214, 296
  alert groups 214
  escalation chains 218
  inbound integration 215, 216
  outbound integration 219, 221
  schedules 221

Grafana Play
  URL 183

Grafana Provider
  reference link 237

grafana_rule_group
  reference link 241

Grafana stack 54
  alerting 54
  debugging 309-311
  deploying 24, 25
  load testing 54
  logs 54
  metrics 54
  preparing 308
  profiling 54
  traces 54
  visualization 54

Grafana stack decisions 309

Grafana stack decisions, factors
  architecture 309
  authentication 309
  data retention 309
  management 309

Grafana Tempo 129

Grafana Terraform provider 237, 238

Grafana tools 21
  Faro 22
  k6 22
  Pyroscope 22

Grafana, using in DevOps life cycle 294
  code phase 294, 295
  deploy phase 296-298
  monitor phase 298, 299
  operate phase 298
  plan phase 300
  release phase 295
  test phase 295

Graphite 121
  architecture 122

groups 287

GZIP compression 120

H

HashiCorp Vault 238

headers 139

Helm 63, 225, 228
  installing 63

histogram metric 36
  examples 36

Homebrew 61
  installing 61
  URL 61

HorizontalPodAutoscaler (HPA) 248

Host Metrics Receiver 154

HTTP API reference
  reference link 240

HTTP endpoint 216

I

incident management 197-199
  after aspects 207
  communication 206
  escalating 205
  identifying 203
incident management, before aspects 199
  noise cancellation 201, 202
  roles and responsibilities 199, 200
  supporting tools 202
Incident Response and
    Management (IRM) 209
Incident Response & Management
    (IRM) users 55
incident response tools 3
incident team communication 206
information Technology Infrastructure
    Library (ITIL) 199
Infrastructure-as-Code (IaC) 225
infrastructure components 45
  compute/bare metal 46
  network devices 46
  power components 46
  Simple Network Management
    Protocol (SNMP) 47
  Syslog 47
infrastructure data technologies 45
Insomnia 255
instant vectors 108
instrumentation
  with libraries 43, 44
internal communication 206, 207
Internet Control Message
    Protocol (ICMP) 204
Internet Engineering Task Force (IETF) 47
IP address matching 90, 93

J

Jaeger 131, 140
Jaeger header 141
JavaScript Object Notation (JSON) 33
Jinja language
  reference link 218
JSON File mode 133
JSON parser 90
JSON Web Token (JWT) 162

K

k3d 254
k6 22, 255, 275
  features 285
  installation link 288
  installing 288
  package formats 288
  running 289
  test life cycle 288
  using, for load testing 283-285
Kubeletstats Receiver 150
Kubernetes
  monitoring, with Grafana 148
Kubernetes Attributes Processor 148-150
Kubernetes Cluster Receiver 152
Kubernetes in Docker (KinD) 63, 254
Kubernetes Objects Receiver 152, 153
Kusto Query Language (KQL) 168
  reference link 168

L

label filter expressions 93
Lambda functions
  instrumenting 280

libraries
for programming languages 44
using, to instrument efficiently 43, 44
Lightweight Directory Access
Protocol (LDAP) 54
Linux
Docker Desktop, installing on 63
load testing 54, 275
k6, using for 283-285
local agent 249, 251
local infrastructure/local platform 294
Locust 255
LogCLI
using 102, 103
log files 28
logfmt parser 91
log formats 28
Common Event Format (CEF) 31
JavaScript Object Notation (JSON) 33
Logfmt 34
Microsoft Windows Event Log 33
NCSA Common Log Format (CLF) 31
overview 30
semi-structured logging 29
structured logging 28, 29
Syslog 34
unstructured logging 29
W3C Extended Log File Format 32
Log ingest 83
log labels 83
log pipeline 88, 89
decolorize 90
IP address matching 90
label filter 91-93
label format 94
line filters 89
line format 94
parsers 90, 91
template functions 93

LogQL 81, 84, 105
features 86-88
query builder 84-86
LogQL Analyzer 102
LogQL metric queries
exploring 95
range vector aggregations 95
log range aggregations 96
logs 7, 8, 54
debugging, from OpenTelemetry
Collector 77, 78
in Loki 69-71
reading, from OpenTelemetry Collector 77
log stream selector 88
Logs with Lambda integration 160, 161
Log Volume Explorer
reference link 312
Loki 20
Loki, Grafana, Tempo, Mimir
(LGTM) stack 3, 19
Loki log data
best practices 101
Loki LogQL Analyzer
reference link 102
Loki Search mode 133
Looker 21
loop 217

M

macOS
Docker Desktop, installing on 62
manual instrumentation 44
Masha Manager 18
goals 18
interactions 18
needs 18
pain points 19

mean time to recovery (MTTR) 10, 297

metric protocols 38

  features 38

  implementing, best practices 39, 40

metric queries 95

metrics 7, 34, 54, 107

  fields 35

  in Prometheus/Mimir 72, 73

metrics query editor 164, 165

metric types 35

  comparing 36

  counter 35

  example data 37

  gauge 35

  histogram 36

  summary 36

MicroK8s 63, 254

microservices architecture 99

microservices mode 252

Microsoft Windows Event Log 33

migrating, from OpenTracing

  reference link 140

Mimir 19, 105, 121

  architecture 123, 124

minikube 63, 254

ML forecasting 223

ML outlier detection 223

ML Sift 223

monitoring 4, 5

Monitoring Query Language (MQL) 163

monolithic mode 252

MQL language specification

  reference link 165

Must have, Should have, Could have, Won't have (MoSCoW) 245

N

National Institute of Standards and Technology (NIST) 199

native language instrumentation 280

Natural Language Processing (NLP) 223

NCSA Common Log Format (CLF) 31

network devices 46

  telemetry examples 46

O

object storage 124

observability 3-5, 11

  pitfalls, avoiding 311, 312

observability, implementing on application or system

  best practices 171

observability platform 300

  architecting 243

  organizations issues, solving with 244

observability systems 201

  components 226, 227

Open Agent Management Protocol (OpAMP)

  reference link 313

OpenAI integration 223

Open Authorization (OAuth) 54

OpenTelemetry

  and Ansible 234

OpenTelemetry Collector 148, 228

  installation, automating 228

  installing 66

  logs, debugging from 77, 78

  logs, reading from 77

OpenTelemetry Collector contributed receivers 303

OpenTelemetry Collector Helm
    chart 228-233
OpenTelemetry Demo application
    access credentials setup 64
    credentials and endpoints, adding 65, 66
    installing 64, 67, 68
    repository, downloading 64
    updating 82, 106, 129, 130
OpenTelemetry Demo Architecture
    reference link 191
OpenTelemetry Demo installation
    troubleshooting 76
OpenTelemetry Kubernetes Operator 233
    advantages 233, 234
OpenTelemetry Protocol (OTLP)
    119, 120, 131, 139, 140
operating expenses (OPEX) 244
operational responsibilities 201
Ophelia Operator 14
    goals 14
    interactions 14
    needs 15
    pain points 15
organizations
    issues, solving with observability
        platform 244
outbound integration 219-221

P

panel plugins for Grafana
    reference link 183
Parca
    reference link 282
parsers 90, 91
pattern parser 91

Pelé Product 16
    goal 17
    interactions 17
    needs 17
    pain points 17
Personally Identifiable
    Information (PII) 309
personas 3
Phlare 22, 276
Podman 62
Postman 255
power components 46
    telemetry examples 47
private data source connect (PDC) 171
production environment 254
profiles 59
profiling 54
profiling data 10
Prometheus 105, 106, 119-121
    architecture 122, 123
Prometheus-compatible systems
    interactions, with metrics data 107
Prometheus modules 303
Prometheus Query Language
    (PromQL) 84, 105, 106
    aggregation 116
    comparison operators 110-114
    data types 108-110
    functions 114-116
    HTTP success rate 117, 118
    metric selection 114
    operators 116
    selection operators 110
    time series selection and operators 115, 116
    writing 114
Prometheus Receiver 153

Prometheus Time Series
    Database (TSDB) 107
proof of concept
    developing 254
pull protocol 119
pull systems 302
push protocol 119
push systems 302
Pyroscope 22, 275
    architecture 282
    data, searching 276-279
    features 276
    overview 276
    profile data, collecting 279, 280
    using, for continuous profiling 276
Pyroscope instrumentation methods
    advantages 281
    drawbacks 281

Q

query editor 132

R

range vector aggregations 95
range vectors 109
rapid application development (RAD) 292
Rate, Errors, Duration (RED) 193, 203
RBAC rollout strategy
    reference link 257
realm 54
real user monitoring (RUM)
    59, 204, 263-265
regular expression parser 91
regular expression (regex) 88
Requests, Errors, and Duration
    (RED) 134, 232

resource platform 303
role-based access controls (RBACs) 243
runbooks 205

S

samples 107
scalars 110
scenarios
    adding, to test to run at scale 287
schedules 221
Scrum 292
security 147
Security Assertion Markup
    Language (SAML) 54
security information and event
    management (SIEM) 243
Security Information and Event
    Management (SIEM) 31
security platforms 303
selection operators, PromQL 110
semi-structured logging 29
Sendmail project 47
server-controlled throttling 120
Service Graph mode 133
service-level agreements (SLAs) 204, 296
service-level indicators (SLIs) 39,
    117, 197, 253, 299
    used, for writing alerts 208, 209
service-level objectives (SLOs) 3,
    39, 163 197, 286, 296, 306
    used, for writing alerts 208, 209
service monitoring
    reference link 165
sibling operator or ~ 137
Simple Network Management Protocol
    (SNMP) 47, 119, 121
simple scalable mode 252

single-node Kubernetes cluster
  installing 63
Site Reliability Engineering (SRE) 199
SLO query builder 165
SNMP Trap information
  example 47
SNMP traps 47, 121
Snowflake 21
software application 294
software-as-a-service (SaaS)
      platform   22, 301
span 8, 40, 41
  fields 41
span ID 8
stack traces 10
State of DevOps report 2023
  reference link 301
StatsD 119, 120
Steven Service 15
  goals 15
  interactions 16
  needs 16
  pain points 16
strategic responsibilities 200
structured logging 28, 29
suggestbot 223
summary metric 36
  examples 36
Support Ticket 54
synthetic monitoring 204
SysAdmin, Audit, Network, and
      Security (SANS) 199
Syslog 34, 47
system architecture
  data collection 248-252
  data production 247, 248
  data storage 252
  data visualization 252

  establishing 246, 247
  for automation tools 253, 254
  for management tools 253, 254
  system failure, handling 252

T

tactical responsibilities 200
tags 287
telemetry 3, 6, 147
  sending, to consumers 258, 259
telemetry collector
  debugging 307, 308
telemetry, from demo application
  exploring 68, 69
template functions 93
templating engine 93
Tempo 20, 130
  architecture 143
  interface 132-134
  read pathway, components 144
  search modes 133
  trace collection 130
  trace fields 130, 131
  write pathway, components 144
Tempo features
  exploring 131
Tempo query language 129
  exploring 135
Terraform 225, 236
  used, for managing alerts 240, 241
  used, for managing dashboards 240, 241
  using, for Grafana Cloud 237
test
  scenarios, adding to run at scale 287
  writing, with checks 286
  writing, with thresholds 286, 287

thresholds
  used, for writing test 286, 287
time series 107
time series database (TSDB) 19, 121, 295
trace 40, 41, 54
  information, reporting 41
  in Tempo 73, 75
Trace Context 140
trace ID 8
TraceQL 135
  >> or descendent operator 137
  > or child operator 137
  ~ or sibling operator 137
  aggregators 138
  data types 138, 139
  mathematical operators 138
  spans, selecting 136
  traces, selecting 136
TraceQL, field types
  attribute fields 135
  intrinsic fields 135
TraceQL mode 133
trace record 40
trace view 132
tracing 40
tracing protocols
  exploring 139
  Jaeger 140
  OTLP 139, 140
  Zipkin 140
troubleshooting, OpenTelemetry
  Demo installation 76
  Grafana credentials, checking 76, 77
  logs, debugging from OpenTelemetry
    Collector 77, 78
  logs, reading from OpenTelemetry
    Collector 77

U

Uber header 141
unstructured logging 29
unwrapped range aggregations 96
  examples 98
User Datagram Protocol (UDP) 120
user personas, of observers 11
  Diego Developer 13, 14
  Masha Manager 18, 19
  Ophelia Operator 14, 15
  Pelé Product 16, 17
  Steven Service 15, 16
Utilization, Saturation, Errors
  (USE) 193, 203

V

Vagrant 255
value types 92
  Bytes 92
  Duration 92
  Number 92
  string 92
Vector 120
virtualization tools 255
virtual user hours (VUh) 55
virtual users (VUs) 275
visualizations 54
  using, in Grafana 183-185
VS Code Extension for Grafana
  reference link 312

W

W3C baggage 143

W3C Extended Log File Format 32

W3C Trace Context 142, 143

Webhooks 219

Web Vitals

  exploring 269-271

  reference link 269

Web Vitals metrics 268

Whisper 122

white-box monitoring 203, 296

whitespace management 218

Windows

  Docker Desktop, installing on 62

Windows Subsystem for Linux
    version 2 (WSL2)

  installing 61

Wireshark 255

World Wide Web Consortium (W3C) 140

write-ahead logs (WALs) 100, 122

WSL 61

Z

Zipkin 131, 140

Zipkin B3 headers 142

‹packt›

Packtpub.com

Subscribe to our online digital library for full access to over 7,000 books and videos, as well as industry leading tools to help you plan your personal development and advance your career. For more information, please visit our website.

Why subscribe?

- Spend less time learning and more time coding with practical eBooks and Videos from over 4,000 industry professionals

- Improve your learning with Skill Plans built especially for you

- Get a free eBook or video every month

- Fully searchable for easy access to vital information

- Copy and paste, print, and bookmark content

Did you know that Packt offers eBook versions of every book published, with PDF and ePub files available? You can upgrade to the eBook version at packtpub.com and as a print book customer, you are entitled to a discount on the eBook copy. Get in touch with us at customercare@packtpub.com for more details.

At www.packtpub.com, you can also read a collection of free technical articles, sign up for a range of free newsletters, and receive exclusive discounts and offers on Packt books and eBooks.

Other Books You May Enjoy

If you enjoyed this book, you may be interested in these other books by Packt:

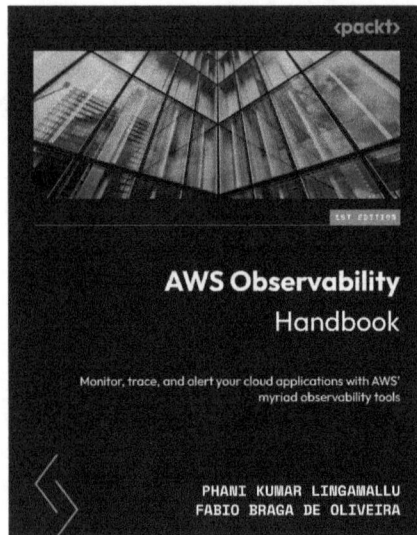

AWS Observability Handbook

Phani Kumar Lingamallu, Fabio Braga de Oliveira

ISBN: 978-1-80461-671-0

- Capture metrics from an EC2 instance and visualize them on a dashboard
- Conduct distributed tracing using AWS X-Ray
- Derive operational metrics and set up alerting using CloudWatch
- Achieve observability of containerized applications in ECS and EKS
- Explore the practical implementation of observability for AWS Lambda
- Observe your applications using Amazon managed Prometheus, Grafana, and OpenSearch services
- Gain insights into operational data using ML services on AWS
- Understand the role of observability in the cloud adoption framework

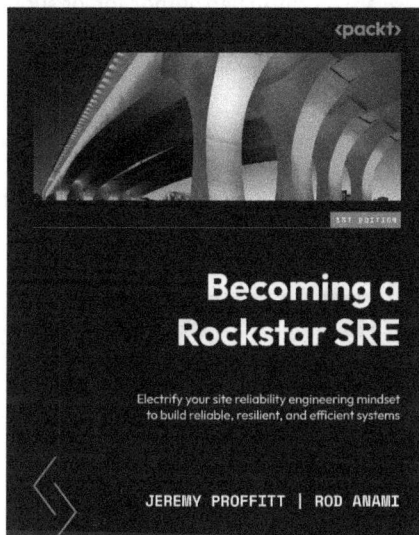

Becoming a Rockstar SRE

Jeremy Proffitt, Rod Anami

ISBN: 978-1-80323-922-4

- Get insights into the SRE role and its evolution, starting from Google's original vision
- Understand the key terms, such as golden signals, SLO, SLI, MTBF, MTTR, and MTTD
- Overcome the challenges in adopting site reliability engineering
- Employ reliable architecture and deployments with serverless, containerization, and release strategies
- Identify monitoring targets and determine observability strategy
- Reduce toil and leverage root cause analysis to enhance efficiency and reliability
- Realize how business decisions can impact quality and reliability

Packt is searching for authors like you

If you're interested in becoming an author for Packt, please visit `authors.packtpub.com` and apply today. We have worked with thousands of developers and tech professionals, just like you, to help them share their insight with the global tech community. You can make a general application, apply for a specific hot topic that we are recruiting an author for, or submit your own idea.

Share Your Thoughts

Now you've finished *Observability with Grafana*, we'd love to hear your thoughts! Scan the QR code below to go straight to the Amazon review page for this book and share your feedback or leave a review on the site that you purchased it from.

https://packt.link/r/1803248009

Your review is important to us and the tech community and will help us make sure we're delivering excellent quality content.

Download a free PDF copy of this book

Thanks for purchasing this book!

Do you like to read on the go but are unable to carry your print books everywhere?

Is your eBook purchase not compatible with the device of your choice?

Don't worry, now with every Packt book you get a DRM-free PDF version of that book at no cost.

Read anywhere, any place, on any device. Search, copy, and paste code from your favorite technical books directly into your application.

The perks don't stop there, you can get exclusive access to discounts, newsletters, and great free content in your inbox daily

Follow these simple steps to get the benefits:

1. Scan the QR code or visit the link below

https://packt.link/free-ebook/9781803248004

1. Submit your proof of purchase
2. That's it! We'll send your free PDF and other benefits to your email directly